Proceedings of the Munich Symposium on Lightweight Design 2021

Jasper Rieser · Felix Endress
Alexander Horoschenkoff · Philipp Höfer
Tobias Dickhut · Markus Zimmermann
Editors

Proceedings of the Munich Symposium on Lightweight Design 2021

Tagungsband zum Münchner Leichtbauseminar 2021

Springer Vieweg

Editors
Jasper Rieser
TUM School of Engineering and Design,
Laboratory for Product Development
and Lightweight Design
Technical University of Munich
Garching, Germany

Felix Endress
TUM School of Engineering and Design,
Laboratory for Product Development
and Lightweight Design
Technical University of Munich
Garching, Germany

Alexander Horoschenkoff
Munich University of Applied Sciences
München, Germany

Philipp Höfer
Universität der Bundeswehr München
Neubiberg, Germany

Tobias Dickhut
Universität der Bundeswehr München
Neubiberg, Germany

Markus Zimmermann
TUM School of Engineering and Design,
Laboratory for Product Development
and Lightweight Design
Technical University of Munich
Garching, Germany

ISBN 978-3-662-65215-2 ISBN 978-3-662-65216-9 (eBook)
https://doi.org/10.1007/978-3-662-65216-9

Editorial Contact: Alexander Grün
This Springer Vieweg imprint is published by the registered company Springer-Verlag GmbH, DE, part of Springer Nature.
The registered company address is: Heidelberger Platz 3, 14197 Berlin, Germany

Contents

About the Editors

Jasper Rieser research is about topology optimization methods with a particular focus on the design for additive manufacturing. Currently, he is a research associate at the Laboratory for Product Development and Lightweight Design at the Technical University of Munich (TUM) from which he also obtained his bachelor's and master's degree in mechanical engineering.

Felix Endress is a research associate at the Laboratory for Product Development and Lightweight Design at the Technical University of Munich (TUM). He investigates product development approaches for metal additive manufacturing, with a special focus on optimization and validation of aerospace structures. Previously, he conducted research in the field of Engineering Design at the University of Cambridge and Friedrich-Alexander-Universität Erlangen-Nürnberg. He holds master's degrees in Mechanical Engineering and Engineering Management.

Alexander Horoschenkoff studied mechanical engineering at TUM and received his Ph.D. from the mechanical engineering department. He started his career at the research center of Messerschmitt-Bölkow-Blohm (MBB) in Ottobrunn. Within the research core team of the DaimlerChrysler AG he was responsible for the mechanical technology field. Since 2001 he has been a professor at the Munich University of Applied Sciences, Department of Mechanical Engineering, Automotive and Aeronautics Engineering and head of the CC "Smart Composites".

Philipp Höfer is a full professor at the Institute of Lightweight Engineering within the Department of Aerospace Engineering at the Universität der Bundeswehr München. After obtaining his Ph.D. in the field of material modelling and continuum mechanics, he has gained extensive experience in the development of aircraft structures at Airbus over many years. His research interests include the conceptual, functional and structural design of lightweight structures and the investigation of their static and dynamic characteristics by analysis and test.

Tobias Dickhut is a full professor of Composite Materials and Technical Mechanics at the Institute of Aeronautical Engineering within the Department of Mechanical Engineering at the Universität der Bundeswehr München. After obtaining his Ph.D. in the field of lightweight construction and structures with fibre-reinforced plastics, he has gained extensive experience in the development of space structures at MT Aerospace over many years. His research interests include the scientific engineering research and design with composites, in particular the issues of lightweight (hybrid) force transmission into highly stressed structural components made of composite materials and the development of tank structures for cryogenic media.

Markus Zimmermann research is about the design and optimization of complex mechanical systems, such as automobiles or robots. Before he became a professor at TUM, he spent 12 years at BMW designing vehicles for crash and vehicle dynamics. His academic training is in Mechanical Engineering with degrees from the Technical University of Berlin (Diplom), the University of Michigan (M.S.E.) and MIT (Ph.D.).

Check for updates

Efficient Computation of Spatial Truss Structures for Design Optimization Approaches Using Tube-Shaped Thin-Walled Composite Beams

Michael Jäger(✉) and Sandro Wartzack

Engineering Design, Friedrich-Alexander-Universität Erlangen-Nürnberg, Martensstraße 9, 91058 Erlangen, Germany
jaeger@mfk.fau.de, wartzack@mfk.fau.de

Abstract. Spatial truss structures are a stiff, economical, and effective lightweight design method, especially when using composites instead of isotropic materials for the struts. An efficient computation of these structures is crucial for optimization approaches during the product design process. The most common method for computing spatial truss structures relies on hinged connections with tension/compression-only struts, which ignores the bending and coupling effects of composite beams. However, especially when using asymmetric laminates, these effects are no longer neglectable. Within commercial finite element tools, the computation of large truss structures - which include these effects - is a very time-consuming process. Particularly for slender, thin-walled beams a large number of solid/shell elements is required. In this paper, an analytical solution of the stiffness matrix for a tube-shaped thin-walled composite beam is provided. It is based on the classical laminate plate theory and Timoshenko's exact solution including shear deformation and coupling effects. By using three-dimensional exact Timoshenko beam elements, the number of degrees of freedom can be reduced significantly while coupling effects are maintained. This results in a remarkably lower computation time especially needed for topology optimization. The results are compared to a commercial finite element tool using both solid and shell elements.

Keywords: Lightweight design · Spatial truss structures · Structural optimization · Thin-walled composite beams · Timoshenko beam

J. Rieser et al. (Eds.): *Proceedings of the Munich Symposium on Lightweight Design 2021*, Proceedings, pp. 1–12, 2023. https://doi.org/10.1007/978-3-662-65216-9_1

1 Motivation

Spatial truss structures are a well-established design method with a high potential for lightweight design [1,2], especially when lightweight materials are used and the topology is optimized during the design process. The most common method for computing truss structures uses hinged connections and tension/compression-only struts [3], which is a very fast and efficient computation method when using isotropic materials e.g. aluminium. However when anisotropic materials are used, this method is no longer suitable, due to the coupled mechanical behaviour of the material. Therefore a coupled model needs to be used for analysis of composite beams like the classical laminate plate theory in combination with the finite element method. Although this is a very powerful tool for calculating composites, it is not advisable for optimizing spatial truss structures due to the enormous computing costs. Classical approaches for truss structure optimization, such as the ground structure method [4,5] and more advanced methods like an adaptive 'member adding' scheme [6], rely on a very large number of members to be calculated. Therefore an efficient method for computing a large number of struts needs to be used, such as a thin-walled composite beam provided by Librescu and Song [7]. Using this beam theory, an analytical solution for the stiffness matrix for a tube-shaped thin-walled composite beam will be derived in this paper, suitable for large scale optimization approaches of spatial truss structures.

2 Thin-Walled Composite Beam Theory

Assumptions

Let h be the wall thickness along the beam assumed constant, let l be any characteristic cross-sectional dimension of the beam (i.e. diameter, height or width) and L its length [7]. In order to apply this thin-walled composite beam theory, the struts must be slender and thin-walled

$$h/l \leq 0.1, \qquad l/L \leq 0.1. \tag{1}$$

A tube-shaped Timoshenko beam and its degrees of freedom (DOFs) at both ends are shown in Fig. 1.

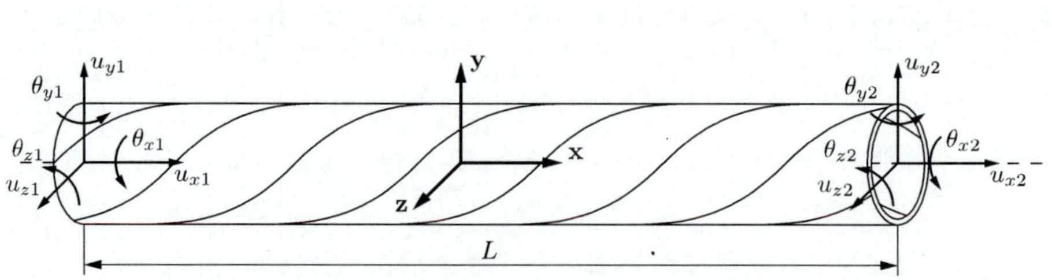

Fig. 1. Timoshenko beam with constant cross section

Additionally, only laminates with a circumferentially uniform stiffness (CUS) configuration [7] (shown in Fig. 2) are considered. Therefore ply layers on opposing sides have to be mirrored

$$\varphi_i(y) = \varphi_i(-y), \qquad \varphi_i(z) = \varphi_i(-z), \tag{2}$$

which is the case for struts manufactured by very common processes like winding, pullwinding, pulltruding or prepreg winding.

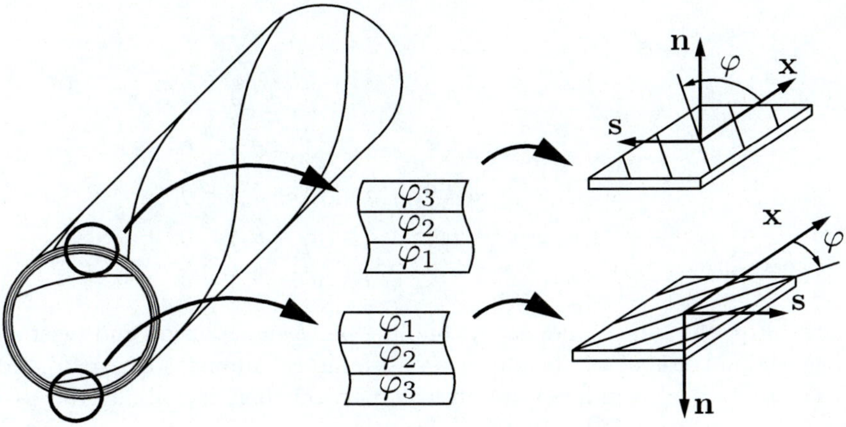

Fig. 2. Circumferentially uniform stiffness (CUS) laminate configuration [7]

Further assumptions for the beam model are, the shape of the cross-section is assumed rigid and remains in its plane, the transverse shear strains are uniform over the beam cross-section [7].

General Beam DAE

Using the symmetry of the CUS laminate configuration, the corresponding cross-sectional stiffness matrix **A** for any closed thin-walled beam has the form

$$\mathbf{A} = \begin{bmatrix} a_{11} & 0 & 0 & a_{17} & 0 & 0 \\ 0 & a_{44} & 0 & 0 & a_{34} & 0 \\ 0 & 0 & a_{55} & 0 & 0 & a_{25} \\ a_{17} & 0 & 0 & a_{77} & 0 & 0 \\ 0 & a_{34} & 0 & 0 & a_{33} & 0 \\ 0 & 0 & a_{25} & 0 & 0 & a_{22} \end{bmatrix}. \tag{3}$$

The equivalent properties of a_{ii} for an isotropic beam are $a_{11} \Leftrightarrow EA$, $a_{44} \Leftrightarrow GA_y$, $a_{55} \Leftrightarrow GA_z$, $a_{77} \Leftrightarrow GI_P$, $a_{33} \Leftrightarrow EI_y$, $a_{22} \Leftrightarrow EI_z$ and $a_{ij} \Leftrightarrow 0$, for $i \neq j$. The derivation **A** will not be described here, for further information please refer to Sect. 4.4-1 in Librescu and Song [7].

Remark: Compared to Librescu and Song, the cross-sectional stiffness matrix $\mathbf{A}$ is permuted to ensure the following displacement vector $\mathbf{u}(x)$ along the centerline of the beam

$$\mathbf{u}(x) = \left[u_x(x)\; u_y(x)\; u_z(x)\; \theta_x(x)\; \theta_y(x)\; \theta_z(x)\right]^{\mathrm{T}}, \quad \text{for } x \in [0, L], \tag{4}$$

with the DOFs ordered equivalent to common finite element analysis software.

The corresponding differential-algebraic system of equations (DAEs) [7] for a Timoshenko beam with symmetric cross section, CUS laminate configuration and a tube-shaped cross section ($a_{44} = a_{55}$, $a_{22} = a_{33}$, $a_{34} = -a_{25}$) is given as

$$a_{11}\, u_x'' + a_{17}\, \theta_x'' = 0, \tag{5}$$

$$a_{17}\, u_x'' + a_{77}\, \theta_x'' = 0, \tag{6}$$

$$-a_{25}\, \theta_y'' + a_{44}\,(u_y'' + \theta_z') = 0, \tag{7}$$

$$a_{25}\, \theta_z'' + a_{44}\,(u_z'' + \theta_y') = 0, \tag{8}$$

$$a_{22}\, \theta_z'' + a_{25}\,(u_z'' + 2\,\theta_y') - a_{44}\,(u_y' + \theta_z) = 0, \tag{9}$$

$$a_{22}\, \theta_y'' - a_{25}\,(u_y'' + 2\,\theta_z') - a_{44}\,(u_z' + \theta_y) = 0. \tag{10}$$

Remarks: Eqs. (5) and (6) indicate a coupling between extension and twist along the longitudinal axis of the beam for an asymmetric laminate ($a_{17} \neq 0$). Equations (7) to (10) also indicate a coupling between bending about the y- and z-axis for an asymmetric laminate ($a_{25} \neq 0$).

Solution of the DAE

Using the boundary conditions

$$\begin{array}{llll} u_x(0) = u_{x1}, & u_x(L) = u_{x2}, & \theta_x(0) = \theta_{x1}, & \theta_x(L) = \theta_{x2}, \\ u_y(0) = u_{y1}, & u_y(L) = u_{y2}, & \theta_y(0) = \theta_{y1}, & \theta_y(L) = \theta_{y2}, \\ u_z(0) = u_{z1}, & u_z(L) = u_{z2}, & \theta_z(0) = -\theta_{z1}, & \theta_z(L) = -\theta_{z2}, \end{array} \tag{11}$$

the DAE can be solved as follows

$$\mathbf{u}(x) = \mathbf{N}(x)\, \mathbf{u_I}, \tag{12}$$

with $\mathbf{u_I}$ representing the node displacement vector (cf. Fig. 1)

$$\mathbf{u_I} = \left[u_{x1}\; u_{y1}\; u_{z1}\; \theta_{x1}\; \theta_{y1}\; \theta_{z1}\; u_{x2}\; u_{y2}\; u_{z2}\; \theta_{x2}\; \theta_{y2}\; \theta_{z2}\right]^{\mathrm{T}}, \tag{13}$$

and the matrix form functions $\mathbf{N}(x) =$

$$\frac{1}{L\,c_1}\begin{bmatrix} n_{11} & 0 & 0 & 0 & 0 & 0 & n_{17} & 0 & 0 & 0 & 0 & 0 \\ 0 & n_{22} & n_{23} & 0 & n_{25} & n_{26} & 0 & n_{28} & -n_{23} & 0 & n_{211} & n_{212} \\ 0 & -n_{23} & n_{22} & 0 & -n_{26} & n_{25} & 0 & n_{23} & n_{28} & 0 & -n_{212} & n_{211} \\ 0 & 0 & 0 & n_{11} & 0 & 0 & 0 & 0 & 0 & n_{17} & 0 & 0 \\ 0 & 0 & n_{53} & 0 & n_{55} & -n_{23} & 0 & 0 & -n_{53} & 0 & n_{511} & n_{23} \\ 0 & n_{53} & 0 & 0 & -n_{23} & -n_{55} & 0 & -n_{53} & 0 & 0 & n_{23} & -n_{511} \end{bmatrix}. \tag{14}$$

The matrix entries of $\mathbf{N}(x)$ are

$$\begin{aligned}
n_{11} &= c_1\,(L-x),\\
n_{17} &= c_1\,x,\\
n_{22} &= 2\,a_{44}^2\,x^3 - 3\,L\,a_{44}^2\,x^2 - 12\,c_2\,x + L\,c_1,\\
n_{23} &= 3\,c_3\,(L-x)\,x,\\
n_{25} &= -c_3\,(2L-x)\,(L-x)\,x,\\
n_{26} &= L\,a_{44}^2\,x^3 + (6\,a_{25}^2 - 2\,c_1 + 18\,c_2)\,x^2 - (6\,a_{25}^2 - c_1 + 6\,c_2)\,L\,x,\\
n_{28} &= L\,c_1 - n_{22},\\
n_{211} &= -c_3\,(L-x)\,(L+x)\,x,\\
n_{212} &= L\,a_{44}^2\,x^3 - (6\,a_{25}^2 + c_1 - 18\,c_2)\,x^2 + 6\,L\,(a_{25}^2 - c_2)\,x,\\
n_{53} &= 6\,a_{44}^2\,(L-x)\,x,\\
n_{55} &= 3\,L\,a_{44}^2\,x^2 - 4\,(c_1 - 9\,c_2)\,x + L\,c_1,\\
n_{511} &= 3\,L\,a_{44}^2x^2 - (2\,c_1 - 36\,c_2)\,x,
\end{aligned} \tag{15}$$

with

$$c_1 = L^2\,a_{44}^2 + 12\,(a_{22}\,a_{44} - a_{25}^2), \qquad c_2 = a_{22}\,a_{44} - a_{25}^2, \qquad c_3 = 2\,a_{25}\,a_{44}. \tag{16}$$

3 Element Stiffness Matrix

The element stiffness matrix $\mathbf{K}$ for a thin-walled tube-shaped beam is obtained by the strain energy $\mathcal{W}$ [8]. With using the same ansatz functions as the displacement functions $\mathbf{u}(x)$ in Eq. (12), an exact Timoshenko beam element is obtained.

$$\mathcal{W} = \frac{1}{2}\int_0^L \boldsymbol{\varepsilon}^{\mathrm{T}}\,\mathbf{A}\,\boldsymbol{\varepsilon}\,dx = \frac{1}{2}\mathbf{u_I}^{\mathrm{T}} \underbrace{\int_0^L \mathbf{B}(x)^{\mathrm{T}}\,\mathbf{A}\,\mathbf{B}(x)\,dx}_{\mathbf{K}}\;\mathbf{u_I}, \tag{17}$$

with $\boldsymbol{\varepsilon}$ being the gradient of the one-dimensional displacement measures represented by $\mathbf{B}(x)\,\mathbf{u_I}$ and $'$ denoting the derivative with respect to x

$$\boldsymbol{\varepsilon} = \begin{bmatrix} u_x' \\ u_y' + \theta_z \\ u_z' + \theta_y \\ \theta_x' \\ \theta_y' \\ \theta_z' \end{bmatrix} = \underbrace{\begin{bmatrix} \mathbf{N}_{1i}' \\ \mathbf{N}_{2i}' + \mathbf{N}_{6i} \\ \mathbf{N}_{3i}' + \mathbf{N}_{5i} \\ \mathbf{N}_{4i}' \\ \mathbf{N}_{5i}' \\ \mathbf{N}_{6i}' \end{bmatrix}}_{\mathbf{B}(x)} \mathbf{u_I} = \mathbf{B}(x)\,\mathbf{u_I}. \tag{18}$$

$\mathbf{B}(x)$ can be expressed using the derivative of the form function matrix $\mathbf{N}(x)$ as

$$\mathbf{B} = -\begin{bmatrix} b_{11} & 0 & 0 & 0 & 0 & 0 & -b_{11} & 0 & 0 & 0 & 0 & 0 \\ 0 & b_{22} & b_{23} & 0 & b_{25} & b_{26} & 0 & -b_{22} & -b_{23} & 0 & b_{211} & b_{212} \\ 0 & -b_{23} & b_{22} & 0 & -b_{26} & b_{25} & 0 & b_{23} & -b_{22} & 0 & -b_{212} & b_{211} \\ 0 & 0 & 0 & b_{11} & 0 & 0 & 0 & 0 & 0 & -b_{11} & 0 & 0 \\ 0 & 0 & b_{53} & 0 & b_{55} & -b_{23} & 0 & 0 & -b_{53} & 0 & b_{511} & b_{23} \\ 0 & b_{53} & 0 & 0 & -b_{23} & -b_{55} & 0 & -b_{53} & 0 & 0 & b_{23} & -b_{511} \end{bmatrix}, \quad (19)$$

with

$$\begin{aligned} b_{11} &= c_1, & b_{22} &= 12\,c_2, \\ b_{23} &= -3\,c_3\,(L-2\,x), & b_{25} &= L\,c_3\,(2\,L-3\,x), \\ b_{26} &= 6\,(L-2\,x)\,a_{25}^2 + 6\,L\,c_2, & b_{211} &= L\,c_3\,(L-3\,x), \\ b_{212} &= -6\,(L-2\,x)\,a_{25}^2 + 6\,L\,c_2, & b_{53} &= -6\,a_{44}^2\,(L-2\,x), \\ b_{55} &= -6\,L\,a_{44}^2\,x + 4\,c_1 - 36\,c_2, & b_{511} &= -6\,L\,a_{44}^2\,x + 2\,c_1 - 36\,c_2. \end{aligned} \quad (20)$$

Using $\mathbf{B}(x)$ and $\mathbf{A}$, the beam stiffness matrix $\mathbf{K}$ can be evaluated as

$$\mathbf{K} = \int_0^L \mathbf{B}(x)^{\mathrm{T}}\,\mathbf{A}\,\mathbf{B}(x)\,dx = \frac{1}{L\,c_1}\begin{bmatrix} \mathbf{K_{11}} & \mathbf{K_{12}} & -\mathbf{K_{11}} & -\mathbf{K_{12}}^{\mathrm{T}} \\ & \mathbf{K_{22}} & -\mathbf{K_{12}}^{\mathrm{T}} & \mathbf{K_{24}} \\ & \text{sym.} & \mathbf{K_{11}} & \mathbf{K_{12}}^{\mathrm{T}} \\ & & & \mathbf{K_{22}} \end{bmatrix} \quad (21)$$

with

$$\mathbf{K_{11}} = \begin{bmatrix} k_{11} & 0 & 0 \\ 0 & k_{22} & 0 \\ 0 & 0 & k_{22} \end{bmatrix}, \quad \mathbf{K_{22}} = \begin{bmatrix} k_{44} & 0 & 0 \\ 0 & k_{55} & 0 \\ 0 & 0 & k_{55} \end{bmatrix},$$
$$\mathbf{K_{12}} = \begin{bmatrix} k_{14} & 0 & 0 \\ 0 & k_{25} & k_{26} \\ 0 & -k_{26} & k_{25} \end{bmatrix}, \quad \mathbf{K_{24}} = \begin{bmatrix} -k_{44} & 0 & 0 \\ 0 & k_{511} & k_{512} \\ 0 & -k_{512} & k_{511} \end{bmatrix}, \quad (22)$$

$$\begin{aligned} k_{11} &= a_{11}\,c_1, & k_{55} &= 4\,c_2\,(L^2\,a_{44} + 3\,a_{22}), & k_{22} &= 12\,c_2\,a_{44}, \\ k_{14} &= a_{17}\,c_1, & k_{511} &= 2\,c_2\,(L^2\,a_{44} - 6\,a_{22}), & k_{25} &= -12\,c_2\,a_{25}, \\ k_{44} &= a_{77}\,c_1, & k_{26} &= 6\,c_2\,L\,a_{44}, & k_{512} &= -12\,c_2\,a_{25}\,L. \end{aligned} \quad (23)$$

Remarks: For vanishing coupling effects ($a_{17} = a_{25} = 0$) i.e. a beam with isotropic material, the beam stiffness matrix is equivalent to the one presented by Karadeniz et al. [8].

4 Numerical Examples

In this section, the beam model with the stiffness matrix in Eq. (21) is compared to finite element analyses performed in ANSYS 2021R1 using both solid

(SOLID185) and shell (SHELL181) elements. First, a single tube model with a highly anisotropic laminate is compared, followed by a single tube model with a quasi symmetric laminate. Finally, a comparison for a spacial truss structure with a total number of 64 tubes is drawn. For all examples, the finite element model with solids is considered to be trusted and used as reference.

For all simulations the following material for each unidirectional ply is used. Young's modulus in fibre direction $E_{\parallel} = 134\,639\,\text{MPa}$, Young's modulus perpendicular to the fibre direction $E_{\perp} = 9894\,\text{MPa}$, shear modulus $G_{\perp\parallel} = 4559\,\text{MPa}$ and Poisson's ratio $\nu_{\perp\parallel} = 0.2630$.

Single Beam - Highly Anisotropic Laminate

For the first comparison, a tube with the following dimensions is used: length $L = 1000\,\text{mm}$, inner diameter $d = 26\,\text{mm}$, wall thickness $h = 2\,\text{mm}$. The laminate is made from 4 layers with a thickness of $t_i = 0.5\,\text{mm}$ each and the corresponding ply-angles (inside to outside) are $\varphi_i = [90°, 12°, 30°, 45°]$. The corresponding finite element discretization is shown in Fig. 3.

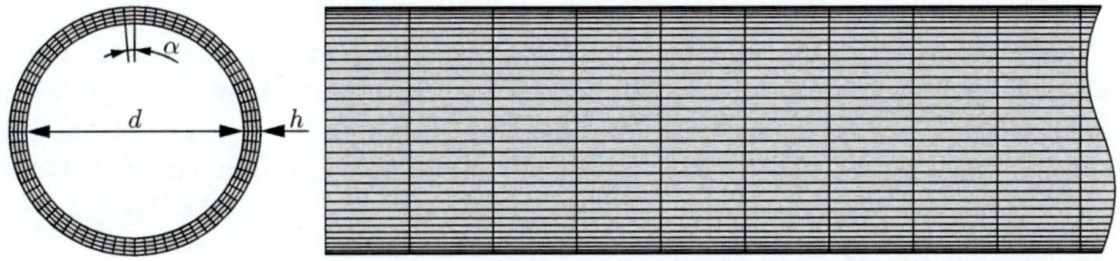

Fig. 3. Finite element discretization of the single tube for solid elements

Along the perimeter, the cross-section is discretized in $\alpha = 5°$ sections resulting in 72 equal elements with an average width of approximately 1.2 mm. Along the length, the tube is discretized in 50 elements with a length of 20 mm each. For the finite element model with solid elements, this gives a total number of 14 400 elements and 56 208 DOFs. The equivalent shell model has 3600 elements and 21 174 DOFs. The FE model is clamped at one end face and load is applied on the opposing end face. For comparison a single beam element based on Eq. (21) is used, the beam is clamped at node one ($u_{x1} = u_{y1} = u_{z1} = \theta_{x1} = \theta_{y1} = \theta_{z1} = 0$) and the load is applied to the second node.

Load Case Tension An axial force of $F_x = 10\,\text{kN}$ is applied, the results are shown in Fig. 4.

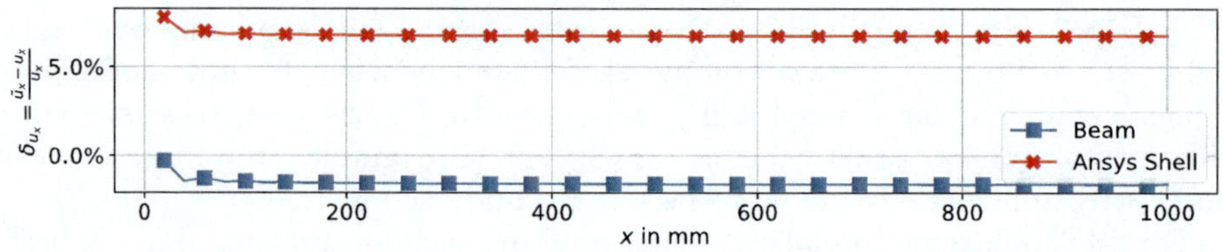

(a) Relative error of displacement $\delta_{u_x} = \frac{\tilde{u}_x - u_x}{u_x}$ relative to the solid model

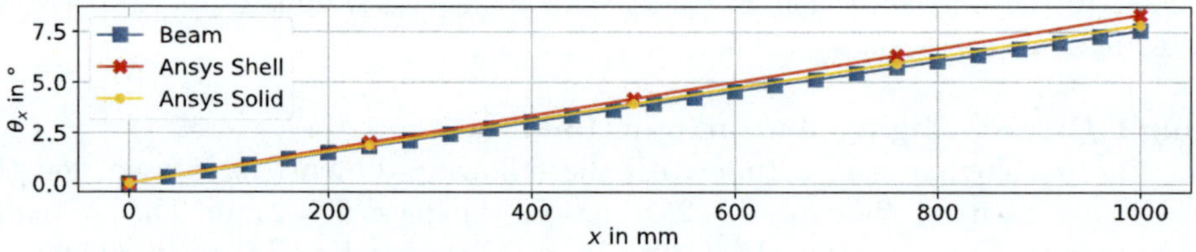

(b) Rotation θ_x caused by coupling of tension and torsion

Fig. 4. Single tube with highly anisotropic laminate under tension $F_x = 10\,\text{kN}$

The displacement u_x and rotation θ_x of the tube under axial tension are well met by the beam model within an error range of less than 2%. The beam model shows a slightly stiffer behaviour than the solid model, but a much smaller deviation than the shell equivalent with an overall less stiff behaviour than the solid model.

Load Case Bending A bending force of $F_y = 50\,\text{N}$ is applied similarly at the end of the tube, the results are shown in Fig. 5.

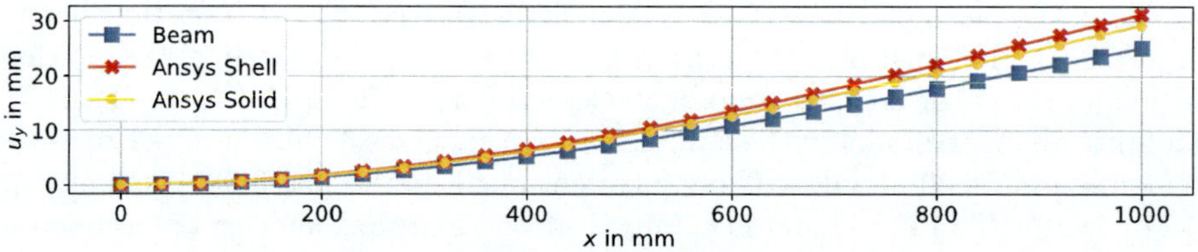

(a) Displacement u_y under bending

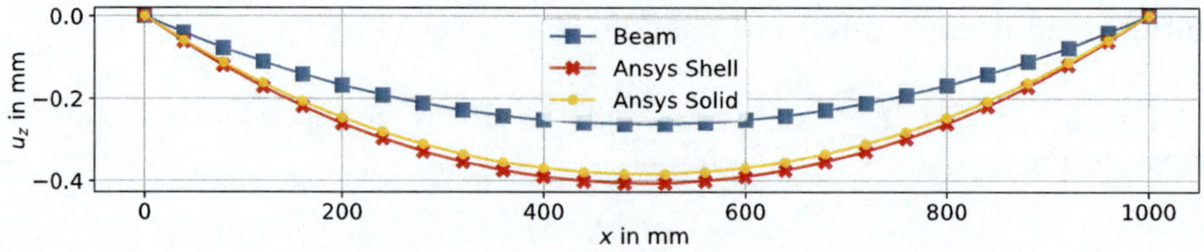

(b) Displacement u_z caused by coupling under bending

Fig. 5. Single tube with highly anisotropic laminate under bending $F_y = 50\,\text{N}$

The beam model still shows a stiffer behaviour than the solid model, resulting in errors of approximately −13.8% in u_y and −31.9% in the coupled u_z direction. The shell model shows a less stiff behaviour than the solid model, with errors of approximately 7% in u_y and 5.6% in the coupled u_z direction.

Single Beam - Slightly Anisotropic Laminate
For this example a more application-oriented laminate made from 6 layers is used. The ply thicknesses are $t_i = [0.2, 0.4, 0.4, 0.2, 0.4, 0.4]$ mm and the corresponding ply-angles (inside to outside) $\varphi_i = [90°, 12°, -12°, 90°, 12°, -12°]$. This results in a quasi symmetric, but slightly anisotropic laminate. All other parameters are retained.

Load Case Tension The displacement u_x and rotation θ_x of the tube under axial tension are shown in Fig. 6. They are very well met by the beam model within an error range of less than 0.6%. In this case the beam model shows a significantly better behaviour in comparison to the shell model with an error above 10% for the displacement and a coupled rotation in the opposite direction.

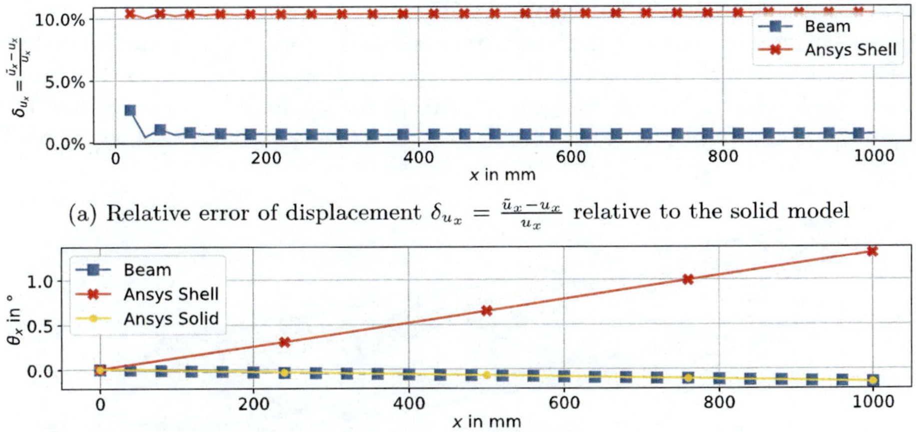

(a) Relative error of displacement $\delta_{u_x} = \frac{\tilde{u}_x - u_x}{u_x}$ relative to the solid model

(b) Rotation θ_x caused by coupling of tension and torsion

Fig. 6. Single tube with slightly anisotropic laminate under tension $F_x = 10\,\text{kN}$

Load Case Bending For the bending load case, the results are shown in Fig. 7. The beam model shows a very accurate result for u_y with an error under 0.5%. The shell model shows similar results compared to the tension load case, with an relative error around 12% for u_y and a false coupling behaviour for the u_z displacement.

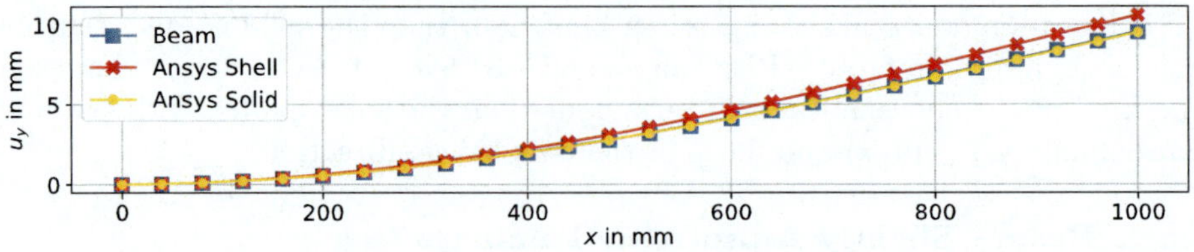

(a) Displacement u_y under bending

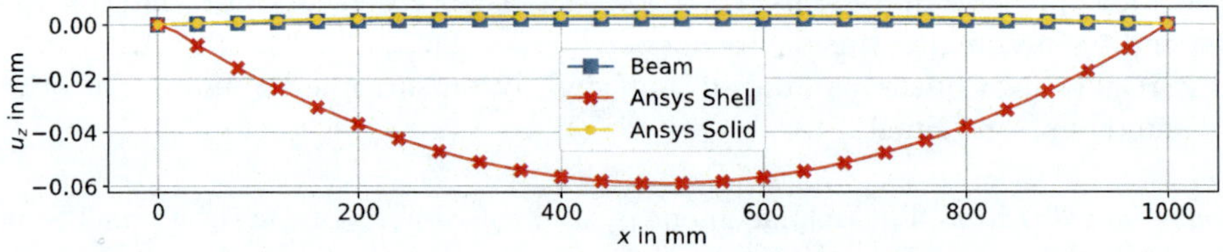

(b) Displacement u_z caused by coupling under bending

Fig. 7. Single tube with slightly anisotropic laminate under bending $F_y = 50\,\mathrm{N}$

Traverse - Slightly Anisotropic Laminate
As final example, a spatial truss structure with 64 tubes made from the slightly anisotropic laminate is compared. The cross section dimensions of the tubes were retained from the examples above, except for the length of each beam. The geometry is shown in Fig. 8, the total size of the traverse is $5000 \times 500 \times 500\,\mathrm{mm}^3$.

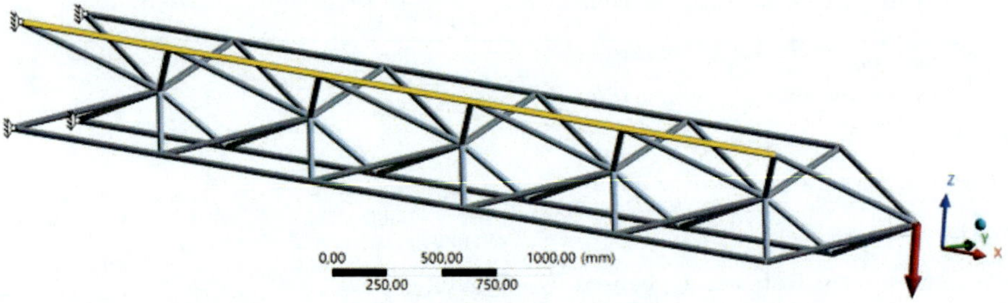

Fig. 8. Traverse model under bending $F_z = -10\,\mathrm{kN}$

The truss structure is clamped at the four nodes on the left side, a bending force of $F_z = -10\,\mathrm{kN}$ is applied at the tip. The displacement is evaluated in Figs. 9 and 10 along the yellow marked path shown in Fig. 8.

The beam model of the truss structure is composed of 64 beam elements with a total number of 156 DOFs. The u_z displacement deviates from the results of the solid model (4 040 376 DOFs) by only 0.6%. In comparison, the shell model's displacement u_z differs by approximately 8% to 9%.

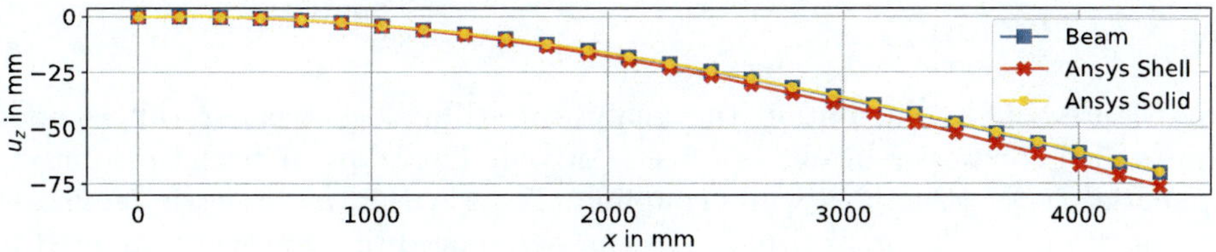

(a) Displacement u_z under bending

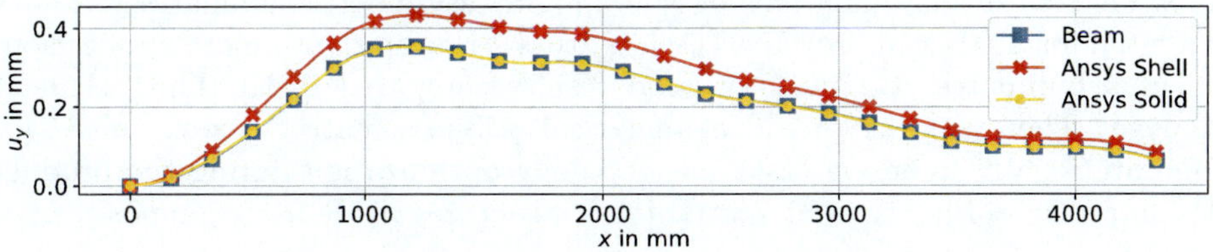

(b) Displacement u_y caused by coupling under bending

Fig. 9. Traverse with slightly anisotropic laminate under bending $F_z = -10\,\text{kN}$

The beam model also mets the coupled u_y displacement very well with an error less than 1%, compared to the shell model (1 047 324 DOFs) with errors up to 35% as shown in Fig. 10b.

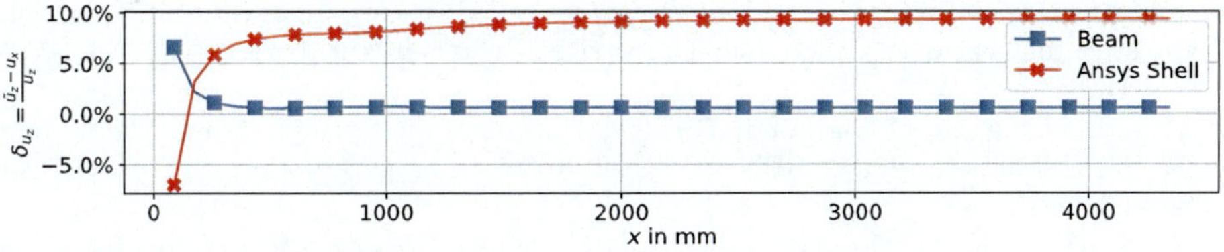

(a) Relative error of displacement $\delta_{u_z} = \frac{\tilde{u}_z - u_z}{u_z}$ relative to the solid model

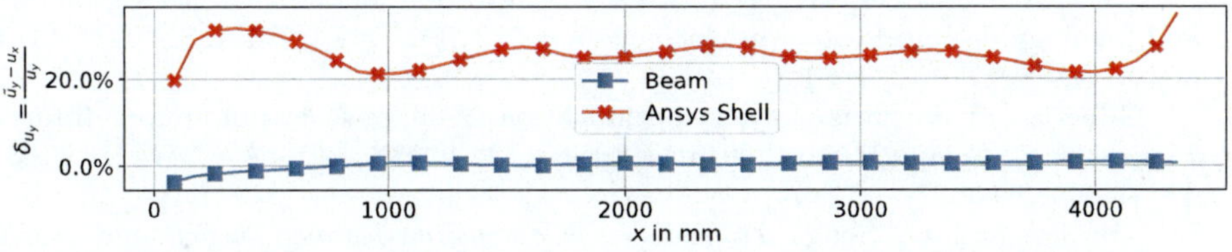

(b) Relative error of displacement $\delta_{u_y} = \frac{\tilde{u}_y - u_y}{u_y}$ relative to the solid model

Fig. 10. Relative error of displacement for a traverse with slightly anisotropic laminate under bending $F_z = -10\,\text{kN}$

Comparing the computation time for this truss, the beam element is orders of magnitude faster. Solving the solid model with Ansys takes 221.02 s (4 CPU cores), preparing the ACP solid model not included (approx. 1200 s). The shell model takes 48.95 s to solve, while the beam model finishes in only 11.86 ms.

5 Conclusion

In the presented contribution the analytical stiffness matrix for tube-shaped thin-walled composite beams has been derived. This allows efficient computing of spatial truss structures while coupling effects within asymmetric laminates are maintained. The number of DOFs can be reduced to a fraction compared to a finite element analysis with solid or shell elements, while errors remain within a reasonable range. Simulations based on highly asymmetric laminates still have potential for further improvement, while truss structures with more application-oriented laminates provide very good results compared to the finite element analysis. Therefore the provided analytical stiffness matrix is well suited for computationally intensive tasks like topology optimization during the product design process of lightweight spatial truss structures made from composites.

Acknowledgments. This Project is supported by the Federal Ministry for Economic Affairs and Energy (BMWi) on the basis of a decision by the German Bundestag.

References

1. H. Pasternak, H.-U. Hoch, and D. Füg. *Stahltragwerke im Industriebau.* EBL-Schweitzer. Ernst, Berlin, 2010. ISBN 9783433600542.
2. D. Klein. *Ein simulationsbasierter Ansatz für die beanspruchungsgerechte Auslegung endlosfaserverstärkter Faserverbundstrukturen.* PhD thesis, Lehrstuhl für Konstruktionstechnik, Friedrich-Alexander-Universität Erlangen-Nürnberg (FAU), VDI Verlag, Düsseldorf, 2017.
3. M. P. Bendsøe and O. Sigmund. *Topology Optimization.* Springer Berlin Heidelberg, Berlin, Heidelberg, 2004. ISBN 978-3-642-07698-5. https://doi.org/10.1007/978-3-662-05086-6.
4. W. S. Dorn, R. E. Gomory, and H. J. Greenberg. Automatic design of optimal structures. In *Journal de Mècanique*, volume 3, pages 25–52, 1964.
5. M. P. Bendsøe, A. Ben-Tal, and J. Zowe. Optimization methods for truss geometry and topology design. *Structural Optimization*, 7(3):141–159, 1994. ISSN 0934-4373. https://doi.org/10.1007/BF01742459.
6. M. Gilbert and A. Tyas. Layout optimization of large-scale pin-jointed frames. *Engineering Computations*, 20:1044–1064, 12 2003. https://doi.org/10.1108/02644400310503017.
7. L. Librescu and O. Song. *Thin-Walled Composite Beams: Theory and Application*, volume 131 of *Solid Mechanics and Its Applications.* Springer-Verlag, Berlin/Heidelberg, 2006. ISBN 1-4020-3457-1. https://doi.org/10.1007/1-4020-4203-5.
8. H. Karadeniz, M. P. Saka, and V. Togan. *Finite Element Analysis of Space Frame Structures*, pages 1–119. Springer London, London, 2013. ISBN 978-1-84996-190-5. https://doi.org/10.1007/978-1-84996-190-5_1.

Substitution of Strain Gauges by Optical Strain Measurement for Standard Test Methods of Composite Specimens and Introduction of a New Biaxial Test-Fixture

Nikolas Korte[1(✉)], Jens Bold[2], and Philipp Höfer[1]

[1] Institute of Lightweight Engineering, Universität der Bundeswehr München, 85577 Neubiberg, Germany
{nikolas.korte,philipp.hoefer}@unibw.de
[2] Boeing Deutschland GmbH, 80805 Munich, Germany
jens.bold@boeing.com

Abstract. Composite parts have a high potential in weight saving for aerospace as well as automotive applications. In the research project RICA, which is funded by the German Federal Ministry for Economic Affairs and Energy, the Universität der Bundeswehr München, the test house BKW Applus+, the composite material manufacturer Teijin and the aircraft manufacturer Boeing are working closely together to enhance material values and allowable by improving material tests to gain more precise values from testing. The project also includes Altair for numerical simulation methods, Vorwerk Autotec for automotive applications, the material manufacturer Henkel and the small and medium enterprise Eckerle for tooling design and manufacturing. Expensive strain measurements will be discussed and modifications to reduce waste and CO_2 in standard testing will be proposed. Complete stress–strain curves are necessary to use advanced failure criteria and material models in simulation to reduce weight in structural parts. However, measurements with the use of strain gauges often cannot provide these complete strain curves until failure, therefore optical strain measurement is investigated for standard test methods. Standard tests were used as baseline and are modified to get more value out of a single test specimen. Additional values needed for modern failure criteria will be discussed and a new Boeing owned bi-axial test-fixture will be presented.

Keywords: Strain measurement · Strain gauge · Optical strain measurement · Digital Image Correlation · Material characterization · Failure criteria · Composite

J. Rieser et al. (Eds.): *Proceedings of the Munich Symposium on Lightweight Design 2021*, Proceedings, pp. 13–24, 2023. https://doi.org/10.1007/978-3-662-65216-9_2

1 Introduction

Material values and allowable are used in todays linear calculations. For more advanced predictions of the load carrying behavior until failure, progressive failure analysis in finite element simulation are used more commonly now. Complete stress–strain curves are needed in all main manufacturing but also loading directions like tension, compression and shear. Strain gauges are used to measure the complete stress–strain curves. In Fig. 1 [1] the distribution of the costs for a complete material card as input for structural calculation is shown. The most expensive part with 58% of the complete costs is related to the strain gauges coming from procurement, installation and cabling. Part of the project is to find a more advanced, cheaper and sustainable method to measure strains in standard tests.

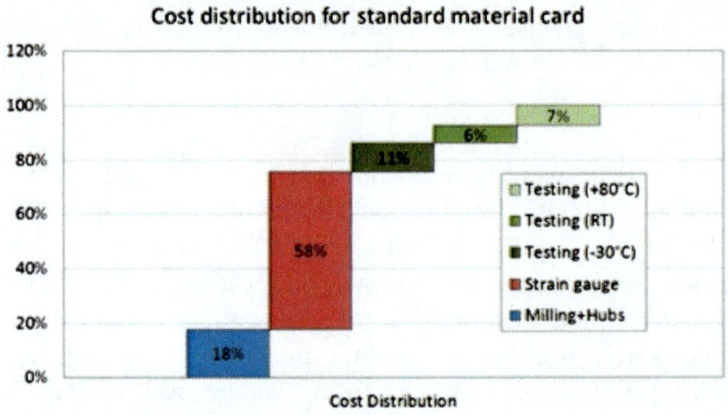

Fig. 1. Cost Distribution for standard material test [1]

In Table 1 a comparison is done with each advantages and challenges for the different systems like: Strain gauges, 1D mechanical measurement, 1D optical measurement, 2D optical measurement (1 camera) and 3D optical measurement (2+ cameras).

Table 1. Comparison of different strain measurement methods

	Strain Gauges	1D Mechanical	1D Optical	2D Optical	3D Optical
Advantage	– more than one direction possible – well known – small	– well known	– no surface preparation, marker required – no damage at failure	– 2D strain distribution – all strains over the complete specimen	– 3D strain distribution – all strains over the complete specimen
Challenge	– expensive – time consuming application – surface preparation	– only one direction – needs space – maybe damages at failure – slipping possible	– only one direction – may not be able to be used at different temperatures	– large amount of data – surface preparation – needs space	– large amount of data – surface preparation – needs space

2 Testing

2.1 Specimen Manufacturing

Composite plates with unidirectional plain carbon fibers and woven glass fibers with an epoxy resin system were used (RTM6/G1157). The composite plates were cut and adhesively bonded with quasi isotropic glass fiber tabs. After curing of the tabs, specimens were cut according to ASTM D3039 [2] with a length of 250 mm, and 25 mm and 15 mm in width. A linear strain gauge in a 3-wire circuit has been applied to the center of each specimen. For strain measurement with Digital Image Correlation (DIC), the surface of the specimens was painted with white acrylic spray. A stochastic speckle-pattern of black dots was sprayed with a paint gun afterwards. The final specimens are shown in Fig. 2a).

2.2 Test-Setup

The specimens were tensile-tested with a ZwickRoell Z150 testing machine with hydraulic clamps. A 150 kN load cell was used. Tests were driven position-controlled with 1 mm/min until specimens' failure. Strain was measured using the following three methods: strain gauge, Digital Image Correlation (DIC) and extensometer. An extensometer (digiclip kurz, ZwickRoell) was used to measure strains up to 0.3% strain and was clamped off above this value to prevent possible destruction of it. Because of the strain gauge and DIC-region of interest, the extensometer had to be clipped above the specimen's center, see Fig. 2b). The strain gauges were connected to a data acquisition device (MGCplus, HBM). Strain measurement with strain gauge and DIC was performed on the same specimen side to avoid different strain values caused by potentially occurred bending moment in the specimen. Through this, a direct comparison between the two methods can be made.

DIC Test-Setup. A Digital Image Correlation system (Q400, Limess GmbH,) with two cameras was used for the optical strain measurement method. The cameras were equipped with macro lenses including spacers, so that the image area was about 35 mm wide and 25 mm high to record the full specimen's width. After focusing the cameras to the specimens' surface, the relative position of one camera to the other was calibrated with a special calibration target. A reference image was recorded of the specimen, clamped one-sided. This image is set as the unloaded condition of the specimen. Also, strain gauges were tare at this moment. A frequency of 1 Hz was used for recording acquisition of the DIC.

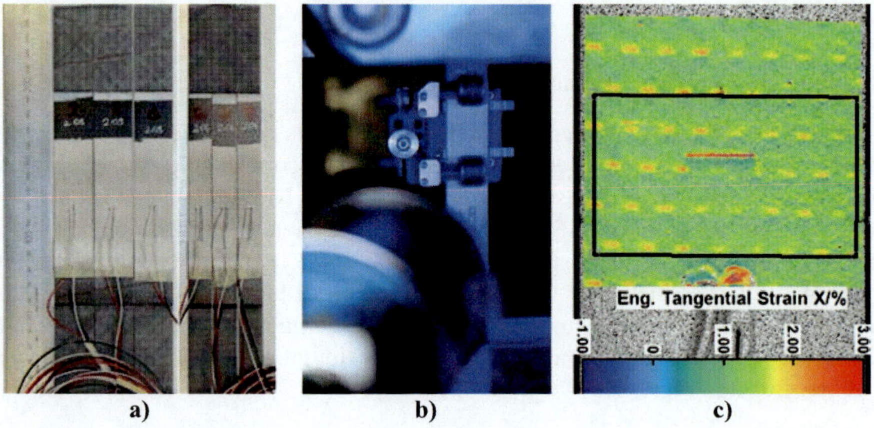

Fig. 2. Test-Setup: a) specimens; b) backside view with extensometer, strain gauge and speckle-pattern; c) DIC-Evaluation field on specimen surface

2.3 Test Results

Evaluation of Strains Using DIC. A coordinate system was set to define the x-direction as the loading direction and the y-direction perpendicular on the specimen's surface to it. For evaluating the strains, an area of the surface was defined, in which the strain is averaged for each time step. This area nearly contains the full specimen width. Figure 2c) shows the evaluation field (black frame) of a specimen.

Stress Strain Curves. The strain measurements of extensometer (Ext), strain gauge (DMS) and DIC are shown in Fig. 3 to compare the different measurement methods on all specimens.

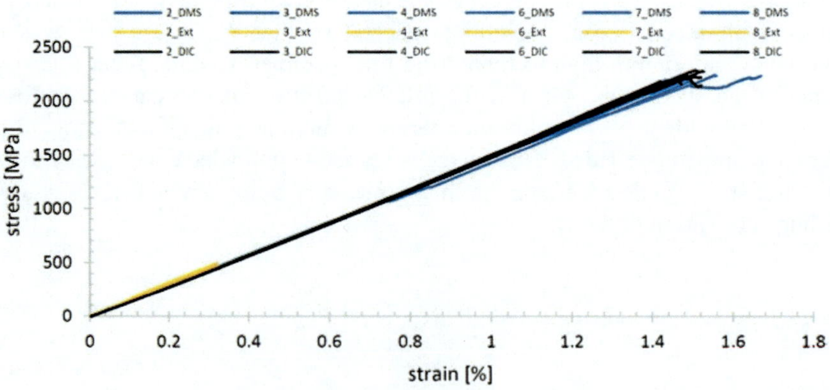

Fig. 3. Stress–strain curves with different strain measurement methods

2.4 Evaluation and Comparison of Strain Measurement Methods

DIC Measurement Scatter. To compare the different strain measurement methods, it is of importance to evaluate the scatter of the DIC measurement. Therefore, the strain in x-direction of a one-sided clamped and unloaded specimen was measured for approximately one minute. Figure 4 shows the calculated strain by the DIC with a computed uncertainty added/subtracted from the actual calculated value. The calculated values show a scatter of about 0.001% strain (in x-direction). Adding and subtracting the uncertainty to/from the actual value (dashed upper/lower line), shows a total scatter of about 0.002% strain (20 microstrains) which is equal to strain measurement requirement in DIN EN ISO 527-1 for Young's modulus estimation based on strain gauges [3].

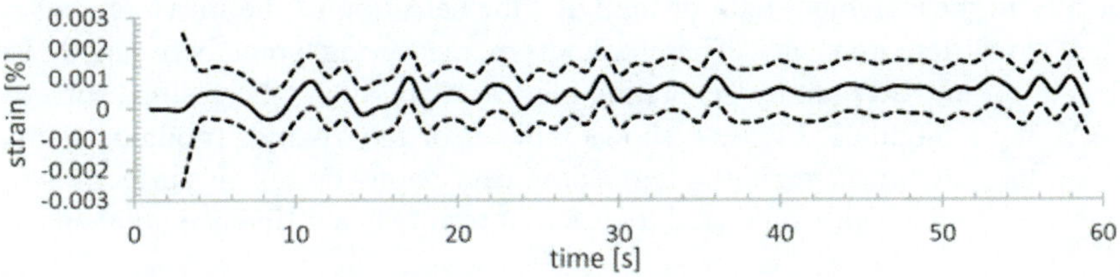

Fig. 4. DIC measurement's scatter of an unloaded specimen

Comparison of DIC and Strain Gauge Values. The calculated strains (in loading direction x) are now compared to the measured strains of the strain gauges of each specimen. Therefore, the absolute deviation was calculated by the difference of gauge strain and DIC strain, following Eq. 1:

$$\Delta\varepsilon = \varepsilon_{DMS} - \varepsilon_{DIC} \quad (1)$$

Figure 5 shows the absolute strain deviation based on strain gauge measurement for every specimen.

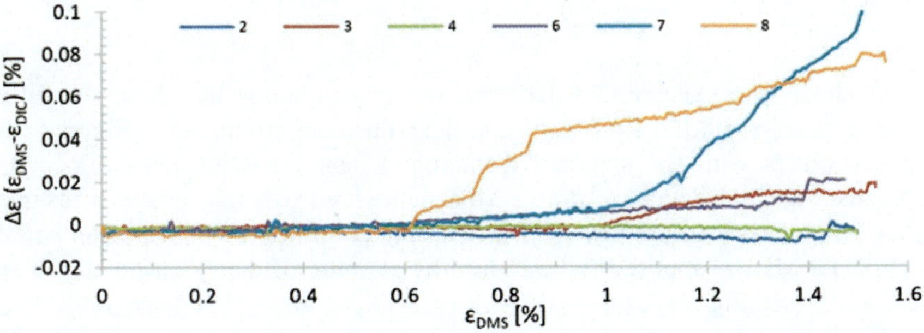

Fig. 5. Absolute deviation [% strain] of measured and calculated strain between strain gauge (DMS) and DIC

Up to a strain of $\varepsilon_{DMS} = 0.6\%$, a maximum absolute deviation of about $\Delta\varepsilon = 0.005\%$ strain is observed for all specimens. The deviation of specimen 8 increases from $\varepsilon_{DMS} > 0.6\%$ strain on, while the other specimens show a smaller increase of deviation up to $\varepsilon_{DMS} = 1\%$ strain, which is about a maximum deviation of $\Delta\varepsilon = 0.008\%$ strain or less. High and sudden deviations (specimens 7 and 8) are assumed to be damages in the strain gauge, or its soldered point.

Comparison of Young's Moduli by Different Strain Methods. Typically, Young's modulus estimation is performed by the difference quotient of stress and strain between 0.1% and 0.3% strain (Eq. 2), according to [2]:

$$E = \frac{\sigma_2 - \sigma_1}{\varepsilon_{x,2} - \varepsilon_{x,1}} \tag{2}$$

Especially in measurement data of the DIC, the selection of the measurement points can lead to different results. Therefore, every measuring point between 0.1% and 0.3% is taken into account by generating a linear regression, whose slope corresponds to the Young's modulus. Figure 6 shows a detail of the Young's modulus estimation based on the DIC-measurement. Here, some data points do not fit perfectly with the linear regression, which would lead to different results in a difference quotient.

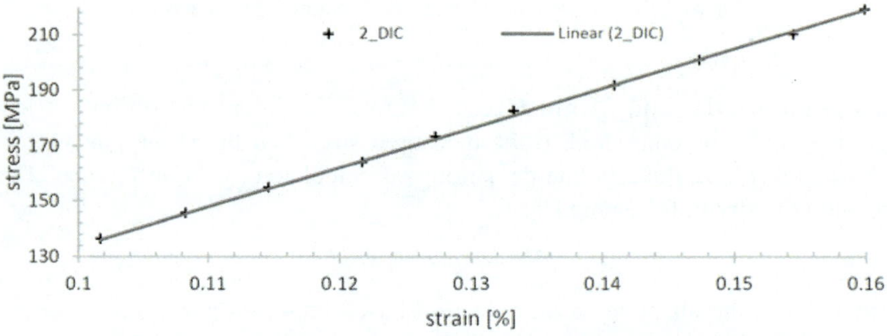

Fig. 6. Stress–strain-curve with linear regression of DIC measurement for Young's modulus estimation

Figure 7 shows the calculated values for Young's modulus based on the different strain measuring methods. It is seen, that the values are comparable. Figure 8 shows the mean values with the standard deviation, where the DIC measurement has a comparable standard deviation (986.1 MPa) compared to strain gauge measurement (1229.8 MPa). The measurement with extensometer shows the highest standard deviation (4891.7 MPa), supposed to be caused by slipping of the extensometer in specimens 2 and 8 (see Fig. 7).

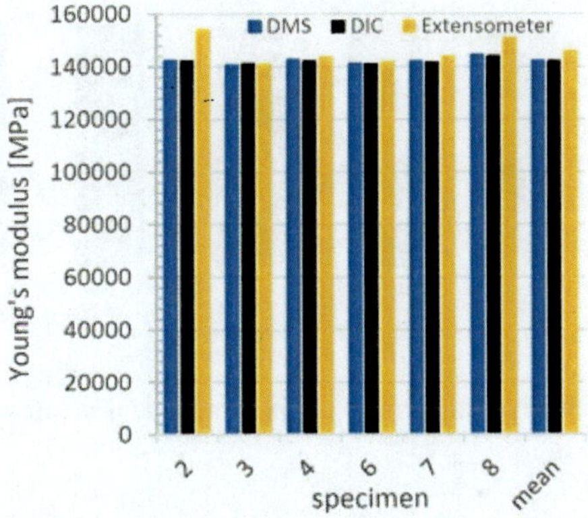

Fig. 7. Calculated Young's moduli by different strain measurement methods for all specimens and mean values

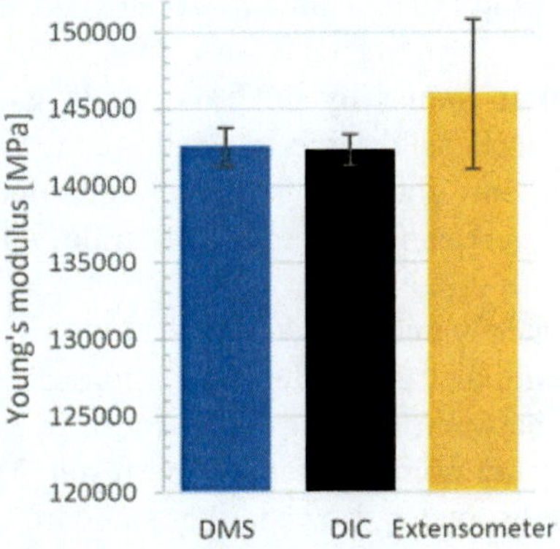

Fig. 8. Mean values and standard deviation of calculated Young's moduli by different strain measurement methods

Beside the evaluated strains in loading (x-) direction, the DIC measurement also offers (y-) strains perpendicular to the loading direction. With these strains, the estimation of the Poisson's ratio is possible, which could not be provided by 1D linear strain gauges. The Poisson's ratio was calculated using Eq. 3 at two measuring points (0.1% and 0.3% strain in loading direction). Figure 9 shows the calculated Poisson's ratio of each specimen and their mean value.

$$\nu = -\frac{\Delta\varepsilon_y}{\Delta\varepsilon_x} = -\frac{\varepsilon_{y,2} - \varepsilon_{y,1}}{\varepsilon_{x,2} - \varepsilon_{x,1}} \tag{3}$$

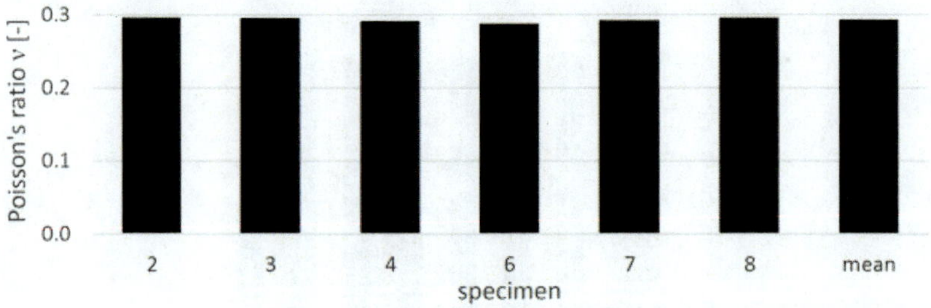

Fig. 9. Calculated Poisson's ratio of specimens and mean value using DIC

2.5 Conclusion

Different strain measurement methods were compared in tensile tests of composite specimens. Uncertainties in the DIC measurement were monitored by conducting a strain measurement with DIC of an unloaded specimen, which showed acceptable results concerning the measurement scatter of about 0.002% strain (with calculation uncertainties).

The comparison of strain values by DIC and strain gauge showed a low deviation until about 0.6% strain. Possibly due to strain gauge rupture, deviation in strain increases for some specimens. Increasing temperature due to lighting of the DIC could also have led to an increasing difference of strain values for strain gauges and DIC.

Comparing the calculated values for Young's modulus, methods based on strain gauges and DIC showed similar values. Here, the approach was to calculate the modulus by linear regression between 0.1 and 0.3% strain to not underly the effect that the selection of the measuring point influences the result. Young's moduli estimation based on values of the extensometer shows higher standard deviation, probably due to extensometer slipping on two specimens.

This work shows the potential using Digital Image Correlation for standard testing. The substitution of strain gauges would lead to less specimen preparation, cost and CO_2 savings, regarding its application and usage. Furthermore, complete stress–strain curves can be recorded, which could be a challenge while using strain gauges because of possible strain gauge rupture during testing.

3 Biaxial Test-Fixture

Interaction values in normal-normal loading are necessary for progressive failure analysis to support testing. They are used in various failure criteria like:

- Composites – unidirectional, woven and non-crimped fabric
  - CUNTZE-BOLD UD & WOVEN Fig. 10 a), b)
  - Tsai-Wu

- non-metallic – resin systems or adhesives
  - von-Mises Yield Criteria Fig. 10 c)
  - CUNTZE-BOLD Adhesives
- metal materials – like titanium, aluminum or steel
  - von-Mises Yield Criteria Fig. 10 c)
  - new CUNTZE-BOLD AM for additive manufactured materials Fig. 10 d)

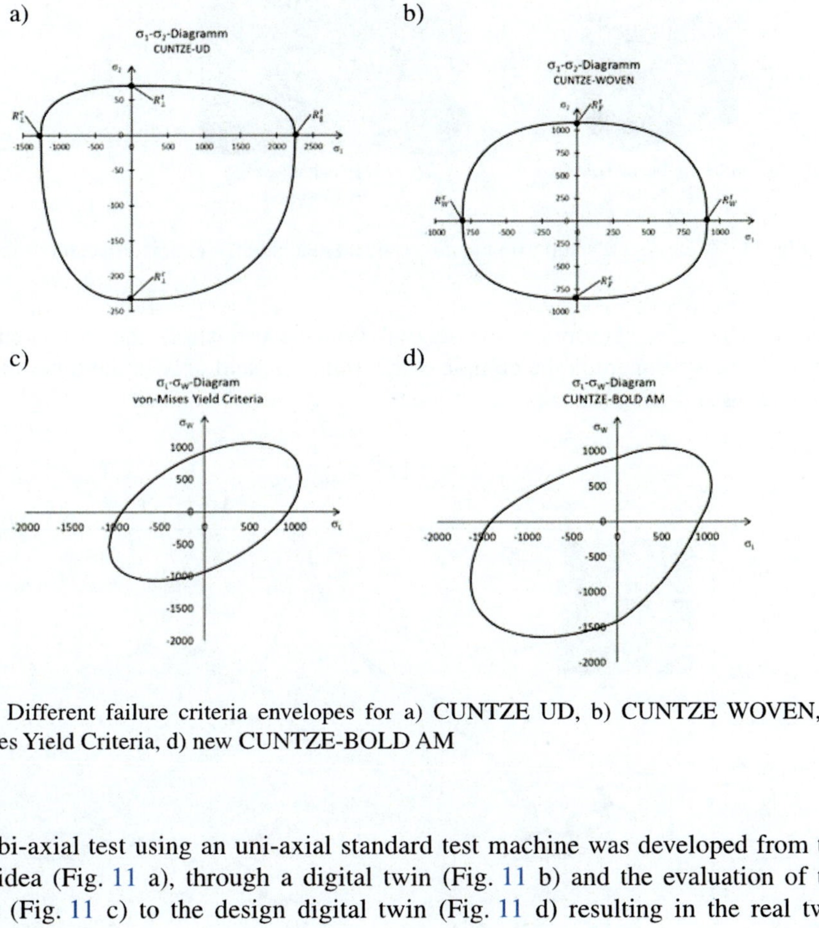

Fig. 10. Different failure criteria envelopes for a) CUNTZE UD, b) CUNTZE WOVEN, c) von-Mises Yield Criteria, d) new CUNTZE-BOLD AM

A new bi-axial test using an uni-axial standard test machine was developed from the design idea (Fig. 11 a), through a digital twin (Fig. 11 b) and the evaluation of the stresses (Fig. 11 c) to the design digital twin (Fig. 11 d) resulting in the real twin (Fig. 11 e).

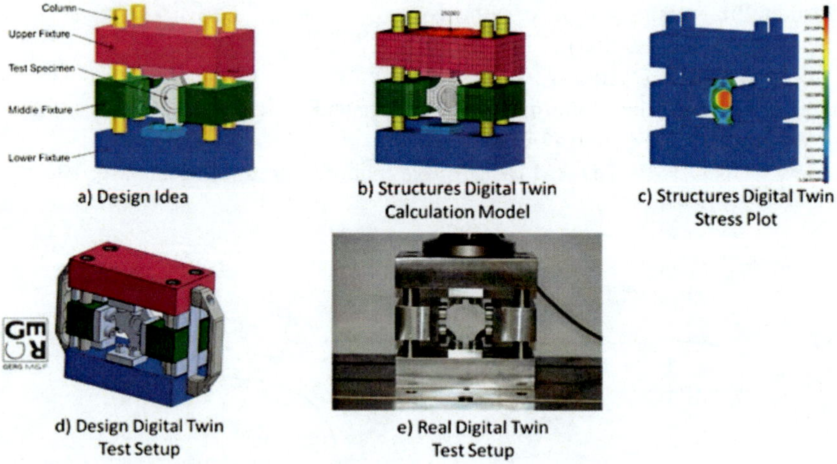

Fig. 11. Bi-axial test setup using a uni-axial test machine (Boeing intellectual property)

The strains were measured using Digital Image Correlation and evaluated in the center of the specimen on the complete area (circle 1) and only in the inner most area (circle 2) as shown in Fig. 12.

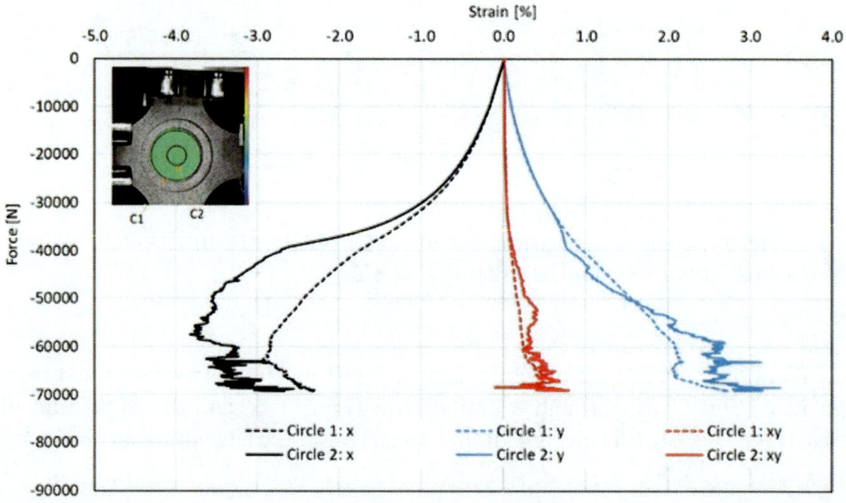

Fig. 12 Strain evaluation using digital image correlation for bi-axial test setup using a uni-axial test machine

The measured strains of the complete test area (C2) were used and the corresponding stiffness was calculated using a bi-linear approach for the non-linear, anisotropic material behavior in the loading direction. These were used to calculate the stresses based on a plane stress state (Eq. 4).

$$\begin{pmatrix} \sigma_W \\ \sigma_T \\ \tau_{WT} \end{pmatrix} = \begin{bmatrix} \frac{E_w(\varepsilon_W)}{1-\nu^2} & \frac{\nu E_W(\varepsilon_W)}{1-\nu^2} & 0 \\ \frac{\nu E_T(\varepsilon_T)}{1-\nu^2} & \frac{E_T(\varepsilon_T)}{1-\nu^2} & 0 \\ 0 & 0 & G_{WT} \end{bmatrix} \begin{pmatrix} \varepsilon_W \\ \varepsilon_T \\ \gamma_{WT} \end{pmatrix} \tag{4}$$

The efforts were calculated and are shown in Fig. 13 a). While the effort for von-Mises yield criteria would predict a value above 1, the new CUNTZE-BOLD AM will lead to an effort around 1. The normal-normal stress is posted in Fig. 13 b) and is very close to the new CUNTZE-BOLD AM non-linear material model and failure criteria envelope.

a) b)

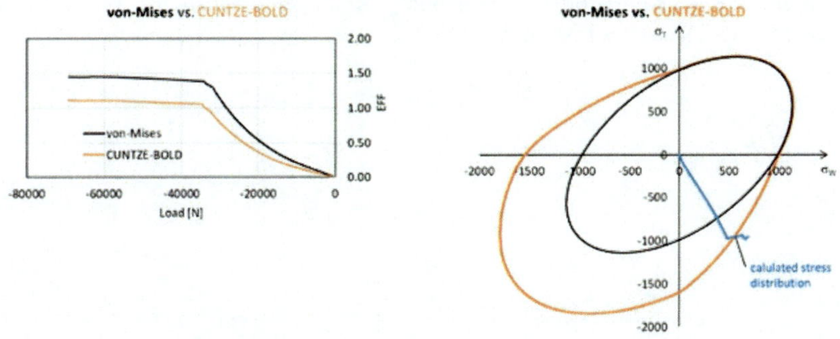

Fig. 13. a) Calculated effort for von-Mises yield criteria and CUNTZE-BOLD AM, b) normal-normal stress posted in failure envelope for von-Mises yield criteria and CUNTZE-BOLD AM

From Fig. 13 one can draw the conclusion that Von-Mises yield criteria would indicate an earlier failure while CUNTZE-BOLD AM is confirmed.

4 Summary and Outlook

The new Boeing owned bi-axial test setup using a uni-axial test machine can be used to evaluate the interaction factor. Further tests with tension loading and different materials are planned.

Acknowledgement. This study is part of the research project RICA, which is funded by the German Federal Ministry for Economic Affairs and Energy, within the program Technologietransferprogramm Leichtbau.

Supported by:

Federal Ministry
for Economic Affairs
and Energy

on the basis of a decision
by the German Bundestag

References

1. Bold, J.: Establishing an In-House process for Composite, Testing – an Overview, Thementag der AG Engineering des Carbon Composite e. V. Augsburg, (2014)
2. ASTM D3039/D3039M – 17 - Test Method for Tensile Properties of Polymer Matrix Composite Materials, ASTM International, United States
3. DIN EN ISO 527-1, Kunststoffe – Bestimmung der Zugeigenschaften – Teil 1: Allgemeine Grundsätze, DIN Deutsches Institut für Normung e. V.

Investigation and Modelling of Machining Processes as Surface Pre-treatment for Structural Adhesive Bonding of CFRP

Jens de Freese[1,2,3](✉)

[1] Wehrwissenschaftliches Institut Für Werk- Und Betriebstoffe (WIWeB), Institutsweg 1, 85435 Erding, Germany
defreese@slv-muenchen.de
[2] Fakultät für Luft und Raumfahrttechnik, Institut für Leichtbau, Universität Der Bundeswehr München, Werner-Heisenberg-Weg 39, 85577 Neubiberg, Germany
[3] Schweißtechnische Lehr- Und Versuchsanstalt (SLV) München, Schachenmeierstraße 37, 80636 Munich, Germany

Abstract. Motivation for the presented work is the increasing application of carbon fiber reinforced polymers (CFRP) in aerospace industry. Here, also increased use of adhesive bonded joints is desired. For surface pre-treatment mechanical processes, such as milling, are applicable very well. For certification of structural adhesive joints pre-treated by milling, further investigations are necessary to build enough knowledge and understanding for safe and reproducible process design.

During this work extensive studies on surface pre-treatment of CFRP by milling were conducted. The resulting surfaces and surface-near laminates were characterized by surface analysis (confocal laser scanning microscopy, scanning electron microscopy, ion-beam cross-section polishing, roughness measurement). Moreover, strength of the bonds manufactured from these surfaces were measured by means of normal force testing. Completing tests with other mechanical pre-treatment processes were performed in parallel. For investigation on the microstructural material separation during milling, a non-linear simulation approach was developed, that could also be extended to sanding/grinding and blasting. The results from simulation were validated with surface analysis of exemplary specimens (sanding/grinding and blasting) and chip root investigations obtained from a quick-stop experiment especially adapted for end milling.

From the combination of surface analysis and simulation a detailed picture of the different effects during material separation could be derived and linked to the machining process input variables. This allows selection of the process parameters, tools and strategies specifically for surface pre-treatment of CFRP, for the first time. Thus, negative effects can be avoided to a large extend and reproducible, high strength adhesive bonding is ensured. The model of microstructural chip formation and the investigation approaches used here are not

J. Rieser et al. (Eds.): *Proceedings of the Munich Symposium on Lightweight Design 2021*, Proceedings, pp. 25–34, 2023. https://doi.org/10.1007/978-3-662-65216-9_3

only applicable for bonding of CFRP but can be transferred to other problems in pre-treatment or machining of fiber reinforced materials in general.

Keywords: Surface pre-treatment for structural adhesive bonding of CFRP · Machining of fiber reinforced plastics · Evaluation of micromechanical chip formation

1 Introduction

During the last decades, composites or fiber reinforced polymers, especially carbon fiber reinforced polymers (CFRP), have become one of the most important materials in lightweight design. With this new class of materials, also new joining technologies are needed. For CFRP, structural adhesive bonding poses different advantages over traditional joining technologies, mainly mechanical fastening such as bolting or riveting.

For the application of adhesive bonding in aerospace applications, reliable surface pre-treatment is necessary to meet high safety standards. Common pre-treatment methods during manufacturing of composite parts are peel-plies or separating foils. For later stage pre-treatment also mechanical processes, such as sanding, are applied. But a lack of knowledge over the complex mechanisms of adhesive bonding of composites together with several incidents lead to a limitation on secondary and non load carrying joints in aerospace applications.

One reason is that pre-treatment of bond lines is often prepared by manual processes. Automation is one approach to overcome limitations regarding size and load. Aim is to replace manual sanding by automated processes, such as automated milling. This gives much more control over material removal and the resulting surface, thus reduce deviations in surface properties and minimize the risk of premature bond line failure.

The work summarized here aims to investigate mechanical pre-treatment of CFRP on a relevant scale for adhesive bonding. Surface phenomena typical for mechanical pre-treatment are characterized systematically, a model for the creation of this surfaces is established by simulation and verified by special designed quick stop experiments and necessary conclusions are drawn to help design mechanical pre-treatment processes for reliable preparation of CFRP bond lines.

2 Experimental

The experiments focus on the pre-treatment of CFRP with end milling, using different parameters and tools. Bonded specimens will be tested in a but joint configuration, as this is most sensitive for damaged CFRP surfaces. Moreover, the pre-treated surfaces are investigated using different surface analysis methods. To understand the influence of the different cutting process parameters, microscale chip formation will be simulated by finite element method (FEM). For validation quick stop experiments

(QS-experiments) are used. On the specimen, produced in the QS-experiment, the "frozen" chip formation can be shown and compared to FEM results.

2.1 Materials and Processes

Materials As substrate material Hexcel Hexply 8552/IM7 UD carbon fiber epoxy laminates were used. The laminates were manufactured from prepreg material and cured following manufacturer specifications. For bonding a composition of the two aerospace standard epoxy adhesives Loctite Hysol EA 9395 and Loctite Hysol EA 9396 at a ratio of 80% to 20% was used, resulting in a paste adhesive of intermediate viscosity. The adhesive was mixed in a centrifugal mixer. Mixing was performed under an absolute pressure of 2000 Pa to avoid air inclusion. Adhesive layer thickness was adjusted by incorporated fillers (50 μm) of the adhesive Hysol EA 9395. Samples were cured at 66 °C for 1 h.

Processes End milling was performed on industrial milling centers allowing high feeds (>10 m/min) and spindle speeds (up to 30,000 1/min). To avoid negative influence on bondline performance, no cooling was used. The tools were different variants of a base tool type HC660BD (Hufschmied Zerspanungssysteme, Bobingen/ Germany). As wear, especially in CFRP cutting the tools were changed after each series of samples. The tool is a tungsten carbide tool with 4 cutting edges, a diameter D_c of 8 mm and different corner geometries (r_e respective f_e). Thus the cutting edge diameter of end cutting edges D_{cO} varies from 2 to 7 mm. Also different other geometry features of the tool such as cutting edge radius r_S, clearance angle α and rake angle γ were investigated by varying this parameters.

During the experiments working engagement a_e was set to 2.38 mm and depth of cut a_p was set to 1 mm. The process parameters feed rate v_f and spindle speeds n are of special interest for all machining processes. Thus, a wide variation of this parameters was investigated resulting in feeds per tooth f_z and cutting speeds on the surface v_{cO} from 0.03 to 0.5 mm and 18.9 to 301.6 mm/min respectively. Corner geometry and feed per tooth result in different chip thicknesses h.

Cleaning of the pre-treated substrates was done manually using lab wipes and solvents (acetone). The procedure was conducted until no further cleaning effect could be noticed by color change of the wipes. To investigate the effect of micro particle residues, that can not be removed by manual cleaning, power ultrasound cleaning (US-cleaning) was used utilizing ultrasonic bath and ultrasonic sonotrode cleaning setups. Further details can be derived from [1–4].

2.2 Testing and Surface Analysis

To measure the performance of the adhesive joints, specimens were tested following DIN EN 15870 [5] using the principle of centrifugal adhesion tensile testing (CATT) utilizing a LUMiFrac Adhesion Analyzer. This destructive testing method allows to apply tensile loads normal to the bond line with low peel or shear stresses [6].

For lower magnification surface analysis and surface roughness measurement, a confocal laser scanning microscope was used. For analysis of the surfaces with higher magnification a field emission scanning electron microscope was used. For detailed investigation of the effects caused by mechanical pre-treatment, cross-sections were prepared. Because of the size of relevant surface phenomena, a very precise cross-section preparation method was needed. Therefore, after conventional cross-section preparation using grinding, a less harmful finishing step was used. This was done by an argon beam cross-section polisher. Here an ion beam removes the last layer of substrate material without creating debris or mechanical loads, resulting in a very flat cross-section without further cracks and scratches. To increase material contrast the cross-sections were etched by low pressure oxygen plasma.

Detailed explanation of testing and surface analysis procedures can be found in [4].

2.3 Simulation and Quick Stop Experiment

To investigate root causes for the surfaces produced by milling pre-treatment, a novel approach for cutting simulation was developed. The simulation represents only a very small fraction of the cutting process, the so called orthogonal cutting model. In this work orthogonal cutting is further simplified by using a 2D-modell. It includes a cutting edge with linearized cutting motion through a small material element of a size of 70×70 μm. The CFRP is modelled in separate fiber, matrix and sizing phases to represent relevant failure behavior according to the investigated phenomena.

Milling is a process where high strain rates up to $1 * 10^6$ 1/s occur [7, 8]. For epoxy based CFRP rate dependent behavior is known [9, 10] resulting mainly from the Matrix material properties at higher strain rates [11–14]. Therefore, time dependent explicite FE modelling was used to incooperate strain rate dependent effects. For the strain rate dependent behavior of the matrix polymer, a constitutive strength model by Cowper and Symonds and a failure model by Johnson and Cook were used together with material constants fit to experimental data from Gerlach et al. [15] and Bagusat [16]. Detailed explanation can be found in [17].

In the quick stop experiments, interruption of the cutting process is forced by laminate layup and cuts and slots in the composite plates. During the milling experiments, the specimens cross-section along with the top sheet is reduced, until the resulting stresses exceed the tensile strength of the material. The cutting force accelerates the specimen away from the tool where the specimen is collected for further investigations. The full experimental procedure can be found in [17]. The chip root investigations are performed with similar ion-beam cross-section preparation and SEM imaging as the surface analysis.

3 Results

3.1 Surface Characterization and Testing of Bonded Specimen

Figure 1 shows the results from CATT testing of CFRP bonds pre-treated with end milling at different parameters and different cleaning procedures. The graph with manual cleaning (Fig. 1 blue bars) shows very high bond strength above 50 MPa but also lower strength at around 35 MPa. Looking at the fracture patterns, two different patterns are visible. In some specimen, large areas of intralaminar failure, where fibers are pulled out of the laminate (black) can be found (Fig. 1 C4). Other series show a lighter, greyish appearance, were fracture occurs directly at the pre-treated surface (Fig. 1 C1). Here the type of fracture can not be clearly classified as adhesive, cohesive or intralaminar failure.

This greyish fracture patterns could indicate an influence of particle residues, resulting in premature failure of the bonds in the layer of micro particles. For sanding pre-treatment, similar effects are already known [18]. Therefore experiments with power ultrasound cleaning methods were performed in order to reduce particles and increase bond strength.

Looking at the results with US-cleaning (Fig. 1 yellow and red bars), different effects are visible. While for some series an increase of strength with US-cleaning was measured (C4 and C6), other series show almost no difference for the different cleaning methods (C7 and C8). For the series C1 and C2 even significantly decreased strength were measured.

This results show, that there are different microscopic effects caused by milling, influencing the macroscopic properties of the CFRP bonds. Further surface characterization and investigation of the microscale chip formation is necessary to understand the different effects.

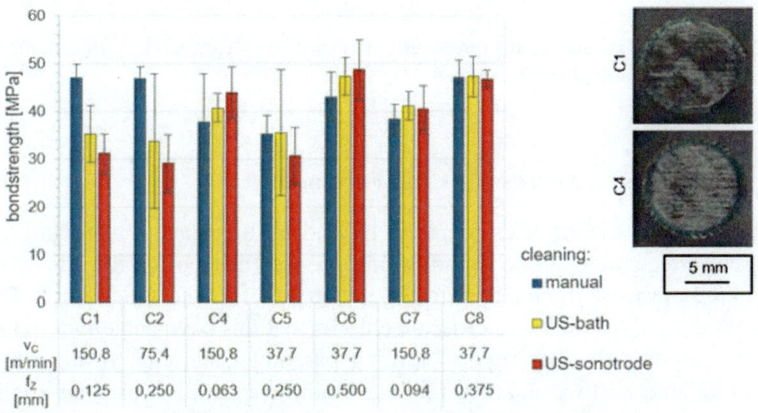

Fig. 1 Bond strength of selected CATT testing series pre-treated with different milling parameters and cleaning methods (left/error bars indicate standard deviation) as well as representative fracture patterns after manual cleaning (right).

The SEM images in Fig. 2 show cross-section of representative CFRP surfaces that where prepared by end milling. In Fig. 2 a the layer of micro particles is very prominent. The particles originate from fiber and polymer matrix material, also loose fibers can be found. Especially the difference between CFRP matrix and loose matrix particles is hardly visible. Due to this layer of micro particles the adhesive can not reach the solid surface of the composite, thus resulting in reduced CATT strength.

This also explains the bright appearance of the fracture patterns of some CATT specimen (Fig. 1 C4). Only by ultrasonic cleaning these particles can be removed. This is why some of the test series showed increase of bond strength due to US-cleaning.

The image shows a second intralaminar flaw typical for end milling pre-treated CFRP (Fig. 2 b). Intralaminar cracks of various length from 1 μm to 30 μm or more. This results in a reduced strength of the CFRP substrate close to the bond line. Even if the adhesive strength of the bond is good, the measured strength of the bond line will be reduced by this intralaminar cracks. This effect is also visible in the irregular form of dark areas on the fracture patterns (Fig. 1 C1). Even more, this effect explains, why not all test series showed positive response to US-cleaning.

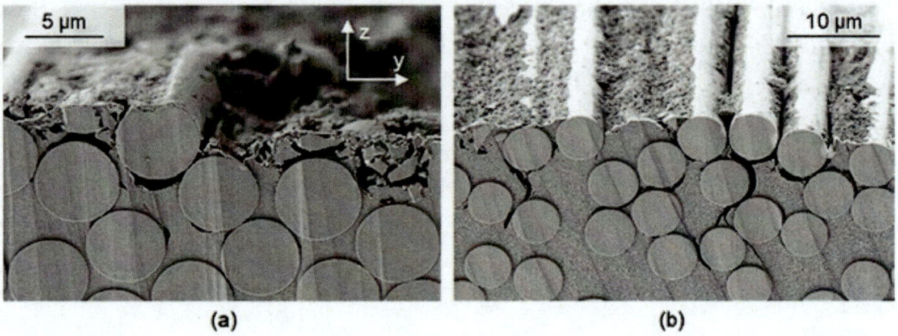

Fig. 2 SEM images of ion-beam prepared cross-section from end milled CFRP surface showing different surface phenomena.

3.2 Investigations on Microscale Chip Formation

Characterization of the surfaces prepared by end milling gave a good impression of the different effects that can be found during pre-treatment of CFRP. Identifying root causes such as process parameters, tool geometry factors etc. is difficult. Therefore, different investigations on the microscale chip formations have been undertaken.

When analyzing the influence of milling parameters on chip formation, the two prominent parameters to look at are feed and spindle speed. For broader applicability, universal parameters to investigate are cutting speed v_C and feed per tooth f_Z.

During simulation a wide range of cutting speeds could be investigated. Figure 3 a and b show a comparison of simulation results with cutting speeds of 75 and 300 m/s. With lower cutting speed, a long crack is induced, producing one large chip, that shears off the CFRP. Smaller cracks under the tool and in the stagnation zone can be seen, resulting from compressive forces, mainly normal/y-force. The crack formed under shear loading mainly travels through solid matrix material, as this is the weakest part of the composite.

With higher cutting speed (Fig. 3 b), shearing produces a network of multiple cracks in front of the tool. This results in a mass of smaller chips rather than one single chip. Crack formation is along the fiber-matrix-interfaces and the cracks travel jumping from fiber to fiber. This is a result of the strain rate hardening effect, modelled for the matrix epoxy. Due to this, strength of the matrix rises and the interface becomes the weakest part of the composite resulting in a brittle failure behavior. The cracks form not only parallel along the shear plane but also downwards into the laminate, thus resulting in a cracked surface for bonding.

The chip roots from the QS-experiment give a very similar picture. In Fig. 3 c chip root produced with 75 m/min cutting speed is depicted. In the primary shear zone as well as in the stagnation and separation zone, only few, very short cracks are visible. In the resulting surface almost no cracks remain. Also micro particles can be seen, which did not follow the chip flow but went underneath the cutting edge and remained on the surface.

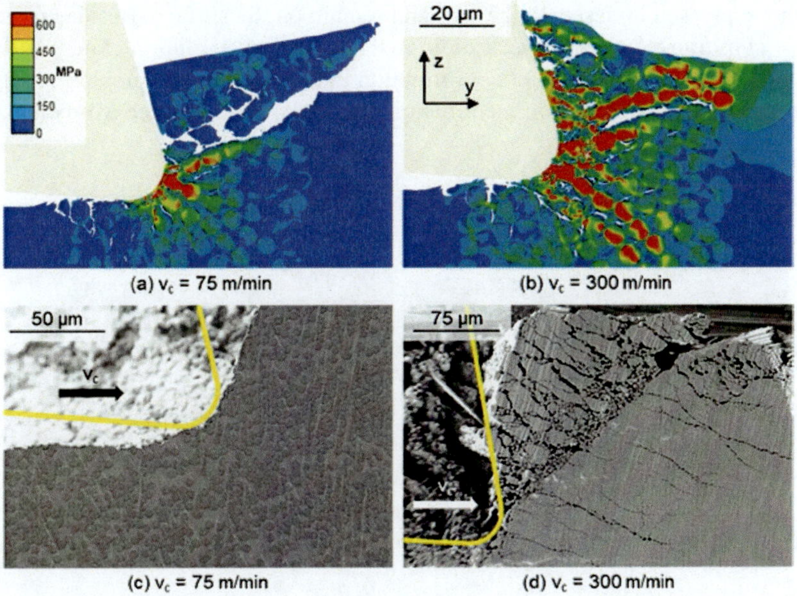

Fig. 3 Comparison of simulation results (von Mises stress) and cross-sections from QS-experiments, both at different cutting speeds.

Figure 3 d is a SEM-images of a chip root produced with a cutting speed of 300 m/min. Here multiple, long (>100 µm) cracks can be found. Also a chip is visible in this chip root. The chip shows even more cracks and also numerous very small fragments and single fibers. The chip root is very similar to the simulation with same cutting speed (Fig. 3 b). When looking to the path of the cracks, it is visible, that, similar to the simulation, the cracks jump from fiber to fiber, thus resulting from fiber-matrix-interface failure.

Another milling parameter to investigate during simulation is the chip thickness h. While in real milling processes chip thickness is a product of feed rate, spindle speed and different geometrical factors, in orthogonal cutting simulation, the chip thickness can directly be investigated. Figure 4 depicts the effect of different chip thicknesses. It is visible, that a certain chip thickness is needed in order to induce shearing and therefore the generation of macroscopic chips. The minimum chip thickness to induce shearing is also called critical chip thickness h_c. Below and above h_c fundamentally different chip formation mechanisms, so called "cutting regimes", can be found.

At high chip thicknesses chip formation is dominated by shearing, thus, this is called "shearing regime". Here, longer cracks and also macroscopic chips are produced (Fig. 4 a). In this regime, large chips and multiple cracks are produced longing deep into the resulting surface.

At very small chip thicknesses the cutting regime is called "rubbing and ploughing". Here microscopic chips are produced, mainly under compressive load. These microchips do not necessarily follow the chip path upwards along the rake face. Many of them are pressed underneath the tool and remain on the surface. Besides ploughing, the poor chip transport also causes additional friction, when chip thickness is too low. Unlike in the shearing regime, crack formation in the rubbing and ploughing regime is limited to a very small area and the remaining material is nearly free from cracks.

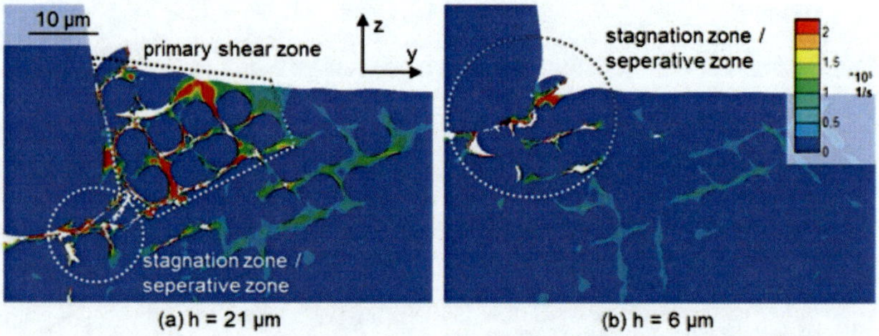

Fig. 4 Images of strain rates during orthogonal cutting of two different chip thicknesses.

The observations on crack formation were verified in QS-experiments. With simulation this observation could be further investigated. The color scale in Fig. 4 depicts strain rate evolution. Simulating different chip thicknesses shows, that strain rate in the cut material is not only influenced by cutting speed, but also by chip thickness. Independent from cutting speed, high strain rates can not be found without shearing. This means, high strain rates and with this brittle fracture is not present at small chip thicknesses or, generally, in the rubbing and ploughing regime.

4 Discussion and Conclusions

With the results from simulation and QS-experiments results from bonding experiments can be explained and parameters for surface pre-treatment can be selected.

When using high cutting speed and medium to high relative chip thickness low and very low (<30 MPa) strength were measured. This is a result of the formation of long cracks under intensive shear and high strain rate. A second combination of parameters with low strength is low cutting speeds together with low relative chip thicknesses. Here different forms of micro damages are produced while the chip formation is in the rubbing and ploughing regime. Both combinations of parameters can not be used for surface pre-treatment.

The formation of micro particle layers was one of the first effects discovered in the experimental results. The micro particle layers are formed in the region of parameters of low to medium chip thicknesses and medium to high cutting speeds. The micro particles significantly reduce bond strength, thus, these parameters are only suitable for surface pre-treatment when combined with powerful cleaning setups such as ultrasonic cleaning.

Only the combination of low cutting speeds (<100 m/min) together with chip thicknesses above critical chip thickness can be used for surface pre-treatment with any cleaning method. Here surface is created with minimum damage and without micro particle layers.

5 Summary

This work presents a very detailed characterization of CFRP surfaces pre-treated with different cutting processes. Moreover, the microscale chip formation has been investigated by a novel simulation approach as well as a new quick stop experiment for end milling of CFRP.

Different damage phenomena occurring during mechanical pre-treatment of CFRP have been identified. Namely, large matrix cracks and interrupted fibers in macroscale and cracked fibers, local fiber delamination and short matrix cracks in microscale were identified. Also the build-up of material layers consisting of micro particles has been found. Different parameters and features could be identified as source of the surface phenomena. With this knowledge, mechanical pre-treatment processes can be optimized for reliable bond line pre-treatment of CFRP for structural adhesive bonding.

References

1. Holtmannspötter, J.: Untersuchungen zum Einsatz von Leistungsultraschall zur Verbesserung klebtechnischer Prozesse. UniBw München, Neubiberg (2010).
2. Holtmannspötter, J., Wetzel, M.; Czarnecki, J. von, Gudladt, H.-J.: How acoustic cavitation can improve adhesion. Ultrasonics 52(7), 905–911 (2012). DOI: https://doi.org/10.1016/j.ultras.2012.02.013
3. Raschendorfer, S.: Behandlung von gefrästen CFK-Oberflächen mittels Primer und Leistungsultraschall zur Steigerung der Festigkeit geklebter Verbindungen. OTH Regensburg, Regensburg (2017).

4. de Freese, J., Holtmannspötter, J., Raschendorfer, S.; Hofmann, T.: End milling of CarbonFiber Reinforced Plastics as surface pretreatment for adhesive bonding – effect of intralaminar damages and particle residues. The Journal of Adhesion 96(12), (2020). DOI: https://doi.org/10.1080/00218464.2018.1557054
5. Deutsches Institut für Normung: DIN EN 15870:2009-08 – Adhesives – Determination of tensile strength of butt joints. Beuth, Berlin (2009).
6. Rietz, U., Lerche, D., Hielscher, S., Beck, U.: Centrifugal Adhesion Testing Technology (CATT) - A Valuable Tool for Strength Determination. Journal of The Adhesion Society of Japan 51(1), 293–297 (2015). DOI: https://doi.org/10.11618/ADHESION.51.293
7. Wedberg, D.: Modelling of high strain rate plasticity and metal cutting. Luleå University of Technology, Luleå (2013).
8. Klocke, F.: Fertigungsverfahren 1: Zerspanung mit geometrisch bestimmter Schneide. 9. Auflage. Springer Vieweg, Berlin, Heidelberg 2018. ISBN: 978-3-662-54207-1
9. Daniel, I. M.; Werner, B. T.; Fenner, J. S.: Strain-rate-dependent failure criteria for composites. Composites Science and Technology, 357–364 (2011). DOI: https://doi.org/10.1016/j.compscitech.2010.11.028
10. Koerber, H., Kuhn, P., Ploeckl, M., Otero, F., Gerbaud, P.-W., Rolfes, R., Camanho, P. P.: Experimental characterization and constitutive modelling of the non-linear stress-strain behavior of unidirectional carbon-epoxy under high strain rate loading. Advance Modeling and Simulation in Engineering Science 5(1), 50 (2018). DOI: https://doi.org/10.1186/s40323-018-0111-x
11. Gilat, A., Goldberg, R. K., Roberts, G. D.: Strain Rate Sensitivity of Epoxy Resin in Tensile and Shear Loading. Journal of Aerospace Engineering 20(2), (2007). DOI: https://doi.org/10.1061/(ASCE)0893-1321(2007)20:2(75)
12. Dandekar, C. R., Shin, Y. C.: Multiphase Finite Element Modeling of Machining Unidirectional Composites: Prediction of Debonding and Fiber Damage. International Journal of Machine Tools and Manufacture 130(5), 537 (2008). DOI: https://doi.org/10.1115/1.2976146
13. Körber, H., Camanho, P. P.: Characterisation of Unidirectional Carbon- Epoxy IM7-8552 in Longitudinal Compression Under High Strain Rates. Proceedings IRF3, (2009). DOI: https://doi.org/10.1051/dymat/2009025
14. Gerlach, R., Pabst, A., Petrinic, N., Hornig, A., Wiegand, J., Siviour, C. R., Hufenbach,W.: The interface between matrix pockets and fibre bundles under impact loading. Composites Science and Technology 69(11–12), 2024–2026 (2009). DOI: https://doi.org/10.1016/j.compscitech.2009.04.021
15. Gerlach, R., Siviour, C. R., Petrinic, N., Wiegand, J.: Experimental characterisation and constitutive modelling of RTM-6 resin under impact loading. Polymer 49(11), 2728–2737 (2008). DOI: https://doi.org/10.1016/j.polymer.2008.04.018
16. Bagusat, F.: Hochdynamische Untersuchungen eines Reinharzes: Bericht E 54/19. Fraunhofer Ernst Mach Institut, Freiburg i. Br. 2019.
17. de Freese, J., Holtmannspötter, J., Höfer, P.: End milling as a surface pretreatment for adhesive bonding of CFRP – new approaches for root cause identification of reduced bond line performance. Proceedia CIRP 85, 227–233 (2019). DOI: https://doi.org/10.1016/j.procir.2019.09.037
18. Holtmannspötter, J., Czarnecki, J., Feucht, F., Wetzel, M., Gudladt, H.-J., Hofmann, T., Meyer, J. C., Niedernhuber, M.: On the Fabrication and Automation of Reliable Bonded Composite Repairs. The Journal of Adhesion 91(1–2), 39–70 (2015). DOI: https://doi.org/10.1080/00218464.2014.896211

In-Mold Coating in Pressing of Natural-Fiber-Reinforced Salt Cores for High-Pressure Die-Casting Applications

Patricia Erhard[1](✉), Dominik Boos[2], and Daniel Günther[1]

[1] Fraunhofer Institute for Casting, Composite and Processing Technology IGCV, Lichtenbergstr. 15, 85748 Garching, Germany
{patricia.erhard,daniel.guenther}@igcv.fraunhofer.de
[2] Chair of Carbon Composites, TUM School of Engineering and Design, Technical University of Munich, Boltzmannstr. 15, 85748 Garching, Germany
dominik.boos@tum.de

Abstract. High-pressure die-casting (HPDC) is known as a highly productive manufacturing process for light metals. To overcome geometrical constraints of die-based production processes, lost cores can be applied to enable internal cavities and recesses. As the molten metal is rapidly shot into the die at high-pressure in HPDC, lost cores need to withstand high mechanical and thermal loads. Salt cores are temperature-stable and can be washed out after casting. Thus, they particularly enable complex internal cavities. Reinforcements with flax fibers significantly enhance the stability of casting cores manufactured by compression molding and thus enable a robust processing with less geometrical restrictions in HPDC. This study presents the method of in-mold coating within compression molding utilized to prevent fiber adhesion, infiltration, and degradation within HPDC. The in-mold coated layers investigated within this study were found to successfully prevent both fiber infiltration and adhesion and enabled high quality surfaces. Moreover, a heat insulation effect was detected. From these indications, a sustainable improvement of the technical cleanliness and even the perspective of a fully recyclable fiber-reinforced salt core is conceivable in light metal applications.

Keywords: Salt cores · Lost cores · Fiber reinforcement · Pressing · Compression molding · High-Pressure Die-Casting · In-Mold Coating · Expendable core · Semi-permanent mold · Fiber-reinforced salt · Natural Fiber · Flax

1 Introduction

High-pressure die-casting (HPDC) is ideally suited for the economical production of light metal components. High tooling costs, low flexibility and limited complexity represent major challenges in die-based production processes like HPDC [1]. Lost

J. Rieser et al. (Eds.): *Proceedings of the Munich Symposium on Lightweight Design 2021*, Proceedings, pp. 35–43, 2023. https://doi.org/10.1007/978-3-662-65216-9_4

cores increase the possible complexity in die-cast parts by enabling internal cavities. However, high injection flows and pressures, combined with high temperatures result in both thermal and mechanical loads that only robust materials can withstand.

In the FASSAD project [2], a robust lost core material based on salt was developed and successfully tested in aluminum HPDC [3] by the partners apppex GmbH, Haas Metallguss GmbH and TUM. In particular, ecologically and economically attractive natural-fiber-reinforced salt cores fabricated by compression molding are investigated. These cores withstand HPDC series conditions at 800 bar cavity pressure, can be washed out with water and are intended to be recycled after casting. However, fibers directly exposed to aluminum melt can be infiltrated. In addition, the unrestrained heat input causes degradation of the natural fiber. The aim of this work is to investigate the suitability of an in-mold coating process for the production of fiber-reinforced salt cores on extreme die-casting process conditions and the development of coating materials therefore. The coating is applied to the mold halves before pressing and backfilled with the fiber-containing core material. Salt is used as the base material for the coating to ensure a material bond between the core and the coating as well as the solubility. Additives are added to provide sealing and insulating effects. The coated fiber-containing salt cores produced in this way are recast in a HPDC process and compared to those without coating. The aim is to investigate the durability of the coating over the core production and casting process, the surface qualities of the cores and cast parts, and the degradation of the fibers.

2 Materials and Methods

2.1 Materials

The base materials investigated in this study (A and B) comprise of 50 wt.-% NaCl of a median particle size of 5.7 μm and 50 wt.-% saturated salt solution, subsequently filled with 10 vol.-% of flax fibers at lengths of 3 and 15 mm. Three compositions of coating materials were developed: composition 1 - a mix of 40.3 wt.-% saturated salt solution, 47.8 wt.-% NaCl, and 11.9 wt.-% bentonite; composition 2 – a mix of 45.5 wt.-% saturated salt solution, 53.7 wt.-% NaCl, and 0.8 wt.-% fumed silica; composition 3 – a mix of 33.9 wt.-% saturated salt solution and 66.1 wt.-% NaCl. A dryly consolidated and milled salt core (C) serves as a benchmark within the study. Table 1 shows an overview of the investigated material compositions. Five specimens of each of the non-coated versions A0, B0, and C were investigated as well as three specimens of each of the coated versions.

2.2 In-Mold Coating and Compression Molding of Natural-Fiber-Reinforced Salt Cores

A cuboid geometry ($100 \times 20 \times 15$ mm^3) is utilized for the experiments. Prior to compression molding, the coating material is applied with a brush to the moving mold and the mold lower cavity surface. The sides of the cavity are omitted as a relocation of the coating material during the manual filling of the cavity is expected. Table 2

Table 1 List of materials

Acronym	Base material	Fiber-reinforcement	Production process	Coating material
A0	NaCl	Flax, 3 mm length	Compression molding	–
A1	NaCl	Flax, 3 mm length	Compression molding	NaCl + bentonite
A2	NaCl	Flax, 3 mm length	Compression molding	NaCl + fumed silica
A3	NaCl	Flax, 3 mm length	Compression molding	NaCl
B0	NaCl	Flax, 15 mm length	Compression molding	–
B1	NaCl	Flax, 15 mm length	Compression molding	NaCl + bentonite
B2	NaCl	Flax, 15 mm length	Compression molding	NaCl + fumed silica
B3	NaCl	Flax, 15 mm length	Compression molding	NaCl
C	NaCl	–	Dry consolidation	–

shows all relevant process parameters within the fabrication process of the salt cores. The compression molding process is conducted using the press station of a RUCKS KV289.00 thermoforming unit, utilizing a tailor-made tool conforming to the geometry described above. The pressurized and temperature-controlled cavity enables the material to consolidate during the dwell time of 10 min. Subsequently to pressing, the salt cores are removed from the cavity and dried in the oven Heratherm OMH750-3P by Thermo Fisher Scientific.

Table 2 Core production parameters

Mold temperature	Cavity pressure	Cavity dwell time	Drying conditions
90 °C	100 MPav	10 min	24 h @ 100 °C, 24 h @ 120 °C

2.3 High-Pressure Die-Casting

The HPDC experiments are conducted on the cold chamber machine Italpress IP 300 SC with a 50 mm piston. The cores are inserted into the casting tool (Fig. 1) and recast. The molten alloy AlSi9Cu3(Fe) is injected directly to the long side of the casting core. Table 3 shows the HPDC process parameters used. A high piston speed and packing pressure is selected to provoke extreme loads on the cores. The cross section ratio between gate and piston is 1:20. The fill time is approximately 0.02 s, the parts are ejected after 7 s and cooled down at room temperature.

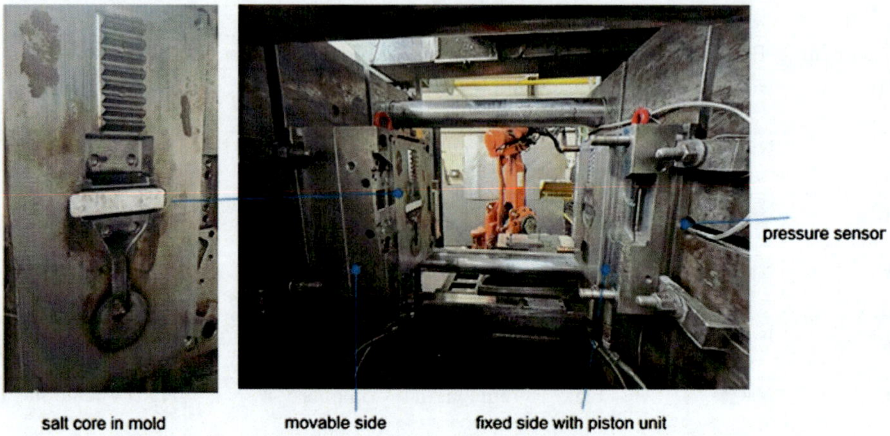

Fig. 1 HPDC experimental setup. Left – a salt core is inserted into the movable side of the mold. Right – the casting tool is installed in the cold chamber machine

Table 3 Parameters in HPDC process

Melt temperature in °C	Piston speed in m/s	Packing pressure setpoint in bar
~630	3.2	1000

2.4 Examination Methods

Photography and Microscopy Photographs document the condition of the salt core after compression molding and drying as well as after HPDC process. Moreover, the cast part is halved by sawing and ground with grade 240 grit paper for microscopic examinations with Olympus SZX10. The photographs and micrographs show the durability of the coating during core fabrication and HPDC process and the degradation of the flax fibers.

Surface Roughness The surface roughness of the cores and cast surface facing the partially coated lower core surface during HPDC is measured using the MarSurf M400 surface measuring instrument and the BFW A 10-45-2/90° probe arm. A total of $n = 6$ measurements are conducted to determine the results.

3 Results & Discussion

3.1 Durability of the In-Mold Coating

Durability during Core Fabrication. The coating materials 1–3 are stable within the core fabrication process (Fig. 2). Only for series A1, two of three specimens' coatings are found chipped off the base material. Since all other coatings are intact, particularly also series B1, it is presumed that an insufficient application of the release agent in the cavity caused this. Thus, the coating materials are considered to interact positively with the base materials.

(a) (b)

Fig. 2 Compression-molded cores of series A – from left to right: A0, A1, A3, A2 (**a**), and series B – from left to right: B0, B1, B3, B2 (**b**)

Durability during HPDC Micrographs (Fig. 3) show the durability of the coatings during HPDC process. No delamination or degradation of the coating material is detected. However, the layer thickness is found to be very uneven. The reason for this is the manual in-mold coating process for lab-scale experiments. However, an automation of the in-mold coating process is intended in industrial application. Closed cracks within the coating are visible. The coated layers were in direct contact with the cast metal but no crack infiltration or relocation can be detected. Thus, shrinkage after HPDC process or pressure after solidification is suspected to induce those cracks. They are therefore rated as uncritical regarding quality of the cast parts but might ease the wash out of the core material.

3.2 Degradation of Flax Fibers

Figure 4 shows different specimens along with the respective shot in the die-casting machine. Only specimens 1 and 2 (corresponding to the scraped parts of material A1, see chapter 3.1) make the insulation effect of the layer visible. The following shots demonstrate an overall degradation. This effect is assumed to be due to the increasing temperatures of the casting tool and melt supply during the first shots. The coating is not able to protect the sides of the cores since the feed of the basic material to the compression molding tool does not allow an in-mold coating here. Thus, the cores are in direct contact with the melt on both sides even for the coated specimens. It is observed that the HPDC parameters used cause full degradation of the flax fibers in production state, affecting the whole specimens' cross-section. However, further investigations are necessary to find out if an in-mold coating applied all around the core will protect the fibers even for the selected harsh conditions.

The insulating effect of coating 1 is visible at specimens 1 and 2. Figure 5 shows exemplary micrographs for the coated and non-coated sections.

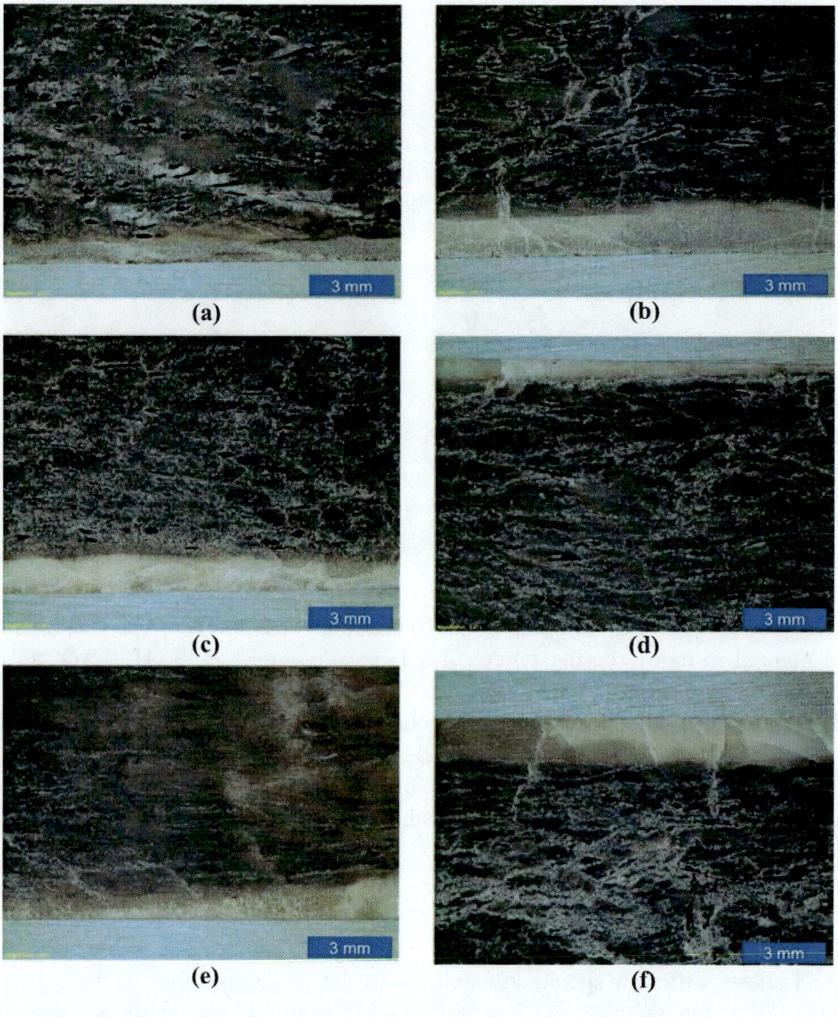

Fig. 3 Micrographs of the coated fiber-reinforced salt cores after HPDC. (**a**) – specimen A1, (**b**) – specimen B1, (**c**) – specimen A2, (**d**) – specimen B2, (**e**) – specimen A3, (**f**) – specimen B3

3.3 Cleanliness and Roughness of Cores and Casting Parts

The surface quality is determined for both the cores and the cast part to identify possible deviations caused by fiber infiltration, degradation or adhesion during HPDC. Figure 6 (a) and (b) shows photographs of non-coated salt cores whereas Fig. 6 (c) and (d) shows the coated versions. The coating enables clean surfaces without fiber infiltrations or adhesion of combustion residues for both 3 mm and 15 mm flax fibers. However, the most significant improvement is achieved when comparing the surfaces

(a) (b)

Fig. 4 Cut and sanded specimens. The numbers indicate the shot in HPDC. During the first shots, degradation only takes place at the non-coated surfaces. The fibers are fully degraded by HPDC in later shots. The fiber-reinforced specimens (**a**) retained their shape despite the high loads except for shot 2 while the pure salt cores broke (**b**)

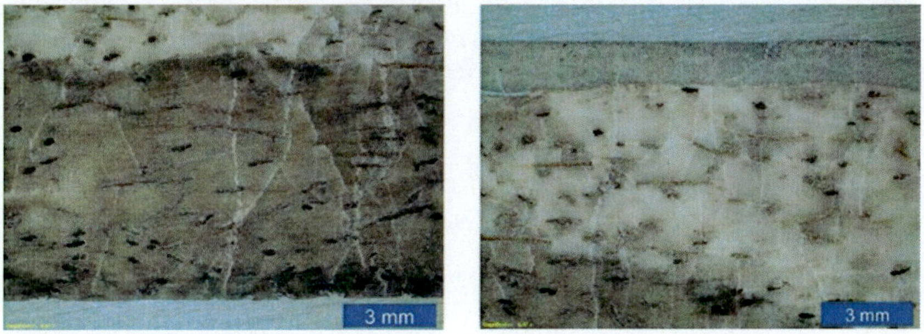

(a) (b)

Fig. 5 Micrographs of the non-coated part of the recast core (**a**) and the coated part of the material composition A1 (**b**)

facing the coated (c) and non-coated (a) 3 mm fiber-reinforced salt core as the coating prevents fiber infiltration.

Surface measurements quantified the roughness of the cores and cast parts. Figure 7 shows the average roughness Ra of each specimens, determined for a total of $n = 6$ measurement points. For every material composition, an average roughness $Ra < 10$ µm was achieved for the compression-molded salt cores. The wide distribution is assigned to the manual application method in connection with condensate formation within the drying chamber. The effect of fiber infiltration as shown in Fig. 6 (a) is confirmed by the results shown in Fig. 7(a). Figure 7(b) shows that no fiber infiltration takes place when using long fibers even without coating, the adhesions visible in Fig. 6 (b) are therefore assigned to combustion residuals that do not affect the surface roughness.

Fig. 6 Surfaces of the cast specimens. (**a**) Specimen A0 corresponding to the non-coated salt cores made with fibers of 3 mm length. Fiber infiltrations and residuals of burnt fibers are visible. (**b**) Specimen B0 corresponding to the non-coated salt cores made with fibers of 15 mm length. Only residuals from burnt fibers are observed, no relevant infiltration with melt takes place. (**c**) Specimen A2 corresponding to a coated salt core made with fibers of 3 mm length. A clean surface was achieved. (**d**) Specimen B1 corresponding to a coated salt core made with fibers of 15 mm length. The coating also prevents the adhesion of fiber residuals to the cast surface

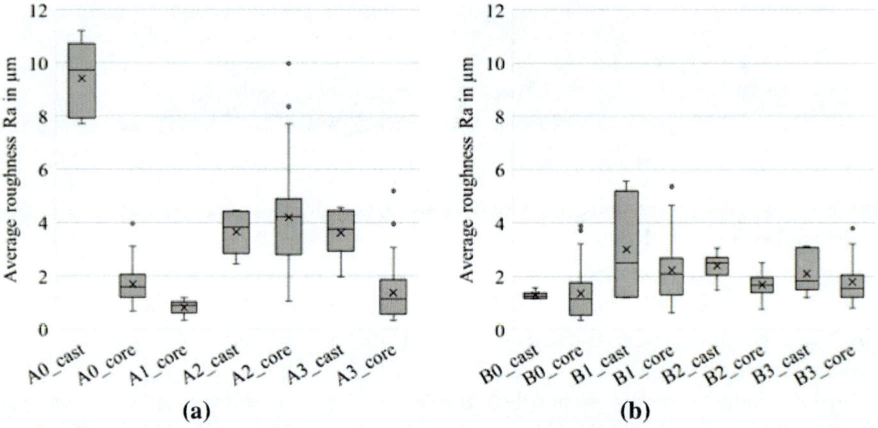

Fig. 7 Surface measurements on the casts compared to the respective cores, (**a**) fiber length 3 mm (A), (**b**) fiber length 15 mm (B). The cast surface of A1 was not measured since the A1 specimens were analyzed for comparing fiber degradation in early and later shots instead

4 Conclusion

This article comprises a study of in-mold coating in compression molding of fiber-reinforced salt cores for high-pressure die-casting. Two previously developed salt suspensions containing flax fibers of 3 and 15 mm [2] were used as well as three coating materials. The study includes investigations on the durability of the coatings over the core production and casting process, the surface qualities of the cores and cast parts, and the insulation of the fibers against heat during HPDC. The absence of delamination and deformation during the whole process indicated an excellent compatibility of the coating and base materials. It was found that the in-mold coating process is suitable for preventing adhesions of burnt fibers to the cast metal. Future investigations may focus on the application parameters of the in-mold coating, geometrical restrictions, and an appropriate feed of the fiber-reinforced salt suspension to the pressing cavity. After casting, the surface roughness of the cast was found to be on the same scale as the core. Moreover, a heat insulating effect of the coating layer was shown. Further investigations are needed to identify the required layer properties for the respectively desired insulating effect dependent on the desired cast geometry, material and properties. At appropriate conditions, a prevention of degradation of fibers to achieve a fully recyclable material system is conceivable in the future.

Overall, fiber-reinforced salt cores show a high potential for applications in HPDC as they can withstand extreme loads and retain removable after casting. Herein it was shown that in combination with an appropriate coating material and process, infiltration of fibers with melt and fiber adhesions to the cast metal can be successfully prevented, enabling a high level of technical cleanliness. Moreover, a prospect to enhance the recyclability of flax fiber-reinforced salt cores can be offered.

References

1. Müller, S., Müller, A., Rothe, F., Dilger, K., Dröder, K.: An Initial Study of a Lightweight Die Casting Die Using a Modular Design Approach. International Journal of Metalcasting 12(4), 870–883 (2018).
2. Boos, D.: Fiber-Reinforced Salt Cores for High-Pressure Casting Processes. In: 26th National SAMPE Symposium. TUM, Munich (2021). https://doi.org/10.13140/RG.2.2.13797.93926
3. Technical University of Munich, Chair of Carbon Composites, https://www.asg.ed.tum.de/lcc/aktuelles/article/erfolgreicher-druckgussversuch-mit-faserverstaerkten-salzkernen/, last accessed 2021/11/16.

Compressibility and Relaxation Characteristics of Bindered Non-crimp-Fabrics Under Temperature and Injection Fluid Influence

Marcel Bender(✉) and Ewald Fauster

Processing of Composites Group, Montanuniversität Leoben, 8700 Leoben, Austria
{marcel.bender,ewald.fauster}@unileoben.ac.at

Abstract. This research examines the compression, as well as short- and long-term relaxation behaviour of bindered textiles at elevated temperature levels. Experiments were conducted on a carbon fibre non-crimp fabric with epoxy resin binder in a specifically designed compressibility test rig. Expanding past research activities at room temperature [1, 2] it was found in series of loading-relaxation-unloading tests, that the test temperature level significantly influences the maximum compaction pressure during the loading stage as well as the pressure characteristics during the relaxation stage [3]. Furthermore, a significant change in the compression behaviour, well below the specified processing temperature of the binder, was found. Also, a proof-of-concept demonstrates the "in-situ"-injection capability of a novel test-rig, reproducing RTM-like conditions in a controlled laboratory environment. The findings of this work are intended to support optimizing preforming and preform handling steps for liquid composite moulding processes.

Keywords: Compaction measurement · Resin-Transfer-Moulding · Preforming

1 Introduction

Beginning with the use of textile as reinforcing structures in polymers, the behaviour of the textile during the saturation process has been of high interest to increase product quality and improve the mechanical behaviour of manufactured components. Especially textile compressibility is of high importance, since it occurs during various stages of the manufacturing process such as preforming, resin-transfer-moulding (RTM) and vacuum-assisted-resin-infusion (VARI). The term compressibility in this

J. Rieser et al. (Eds.): *Proceedings of the Munich Symposium on Lightweight Design 2021*, Proceedings, pp. 44–58, 2023. https://doi.org/10.1007/978-3-662-65216-9_5

case means the textile behaviour during mechanical compression (preforming, mould closing, bagging, etc.) or hydrodynamic compression (resin injection, fluid infusion, wet pressing etc.). Therefore, gaining knowledge of the compaction behaviour and applying it to manufacturing processes increases process control, improves process performance and reduces production waste.

To achieve this, different test set-ups and methods were designed to better understand the process parameters influencing textile compaction behaviour. It was shown that multiple parameters must be considered to maximize the process potential. For example, [1, 4, 5] showed a strong influence of the number of compaction cycles on the peak pressure during compaction and resulting pressure at the end of the relaxation period. In addition, a significant difference in wet and dry compaction behaviour was reported in various works [1, 3, 6], which was attributed to a lubrication effect of the measuring fluid, allowing for easier fibre repositioning. Another parameter controlling the relaxation behaviour in particular, is the amount of binder applied to the textile for preforming purposes. Furthermore, Wu et al. [8] showed that with increasing binder amount on a textile, the necessary compaction pressure at room temperature increases as well. At temperature levels above binder activation temperature, however, this relationship is reversed and increasing binder amount results in lower compaction pressure.

This shows that the relationship between process parameters and textile compaction pressure is diverse. Nevertheless, all of these results share the fact that each parameter was analysed independently and different test set up and methodologies were utilized. As a result, a new test rig was designed, allowing for multiple parameters to be measured independently, as well as in dependence on each other.

2 Compressibility Measurement

2.1 Novel Test Rig

As the 2021 benchmark on textile compaction showed, a large number of different compressibility test rigs are available in the scientific community [9]. Nonetheless, all of these were designed independently and to fit specific testing capabilities. However, to compare the influences of more than one specific testing parameter on the compression behaviour of bindered fabrics one test rig supporting multi parameter testing capabilities is necessary. For this, a novel test rig, displayed in Fig. 1 was designed, aiming to reproduce RTM-like conditions in a controlled laboratory environment.

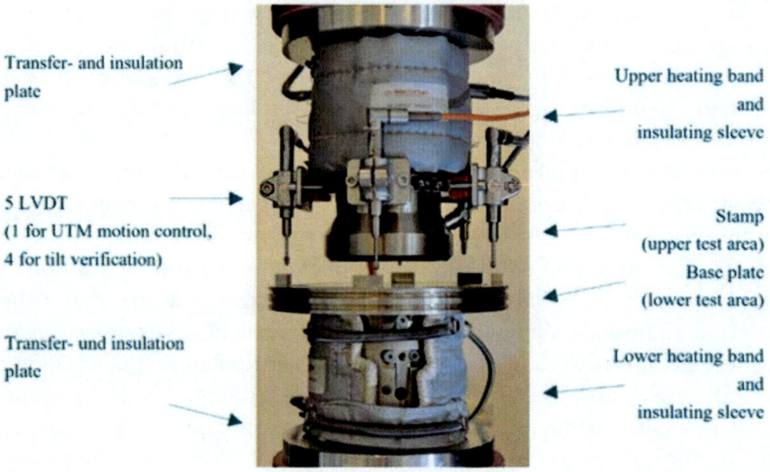

Fig. 1 Novel test rig designed for multi parameter testing capabilities

Since in RTM-like processes, compression and wet pressing occur at elevated temperature, heating bands are situated above and below the test areas allowing a homogeneous temperature distribution across both test areas of 100 mm diameter. In order to avoid unwanted effects of the thermal test rig expansion on the load cell readings a heating solution was designed based on frequency modulation, i.e. adding randomized to periodic heating pulses. The periodic heating pulses carry the load of test area heating, while the randomized pulses break periodicity and stabilize the thermal expansion of the test rig. Figure 2 shows an example set of such heating pulses.

With this method it is possible to give a systematic heating pulse with a maximum frequency of 20 Hz and hold it for a minimum duration of 0.05 s, which results from a cycle time of the heating control process of 20 Hz. Additionally, the duration and occurrence of the randomized heating pulses can be controlled individually, but with the same frequency of 20 Hz.

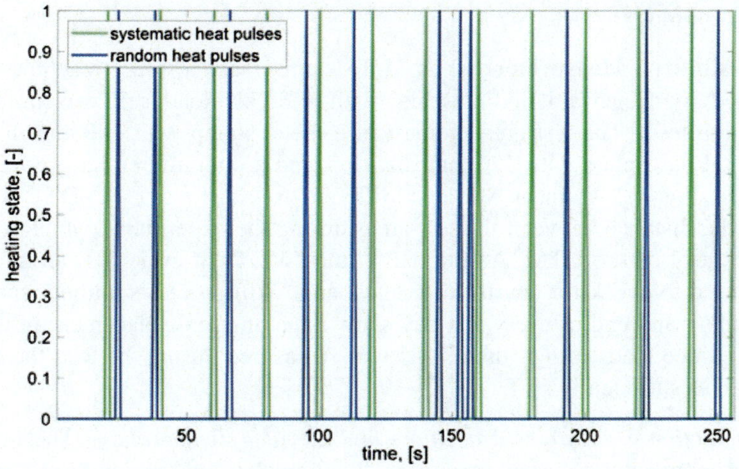

Fig. 2 Exemplary chosen set of systematic pulses (green) and randomized pulses (blue) at an update frequency of 20 Hz. Combined these two signals result in the overall heating pulse signal

Furthermore, the test rig allows for fluid injection into the centre of the specimen through the base plate, to replicate a process step in which the compressed, dry textile is saturated by resin at elevated temperatures.

While test area heating and fluid injection are permanently integrated into the test rig, they can be used independently of each other increasing the testing capabilities, while at the same time avoiding measurement errors resulting from mechanical modifications.

2.2 Textile

For the dry compaction tests at elevated temperatures a Hexcel X-C-PB non-crimp-fabric (NCF) [10] was used. The carbon fibre NCF consists of two ±45° layers with 12 k rovings and has 2.7% Hexion Epikote Resin 05311 applied on the top layer [11]. The areal weight of the textile is 555 g/m^2. The textile was cut to patches of 120 mm × 120 mm using a CNC-cutting machine and stacked to a uniform 7-layer stack. The stacks have an average areal weight of 555.21 g/m^2 with a standard deviation of 1.9 g/m^2 and thus can be seen as uniform.

For the dry compaction measurements with "in-situ" saturation a Hexcel 1202 woven glass fibre (GF) fabric [12] with a fibre areal weight of 290 g/m^2 was used. The textile was cut to patches of 120 mm × 120 mm using a CNC-cutting machine and stacked to a uniform 14-layer stack.

2.3 Methodology

Compressibility Measurements. A UTS universal testing machine (UTM), refurbished by Hegewald und Peschke with a 30 kN load cell was used for the compression tests. The test areas consist of a metal stamp with 100 mm in diameter and a metal base plate with 120 mm diameter and a circumferential groove to drain fluids (see Fig. 1). A linear variable differential transformer (LVDT) is used to measure the distance between the test areas during the experiment, which is used for motion control of the UTM. Additionally, four LVDTs at every 90° around the test area are used to check for any tilt of the test area. Stiffness tests without specimen at the beginning and end of every test day show a maximum displacement of 0.005 mm at 20 kN. Since loads of less than 3 kN were measured during the test, the test rig is considered as sufficiently stiff.

Dry Compaction Measurement at Room and Elevated Temperature. The test regime for the dry compaction measurements at room and elevated temperatures consisted of heating-loading-relaxation-unloading stages, as displayed in Fig. 3. The initial heating phase of five minutes, during which the gap is closed, until the specimen comes into contact (< 25 N) with both test areas, allows the specimen to consistently reach test temperature. During the following loading phase, the specimen was compressed with a closing speed of 1 mm/s, to a calculated fibre volume fraction (FVF) of 0.58, resulting in the compressed specimen thickness of 3.721 mm. After reaching the set specimen thickness a relaxation period of one hour, respectively 12 h for the long-term measurements, begins, during which the crosshead position is held constant and

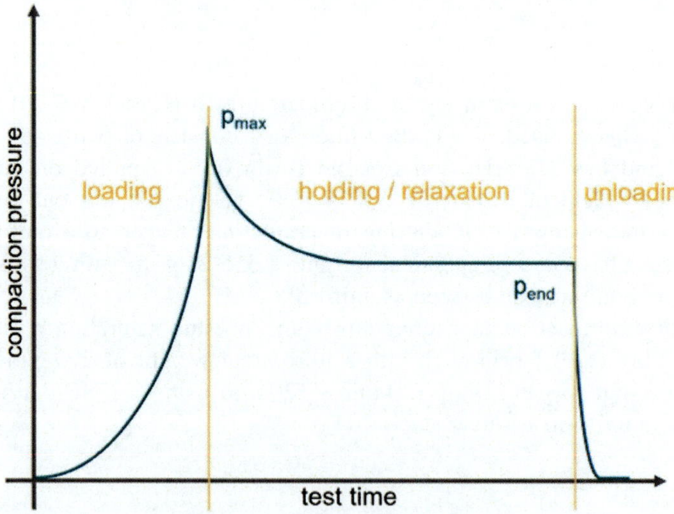

Fig. 3 Schematic of a dry compaction measurement. An optional heating phase for measurements above room temperature starts before the actual test time

the pressure decreases. Finally, in the unloading stage, the test areas are moved apart with an opening speed of 1 mm/s.

Combined Dry and Wet Compaction Measurements (in-situ Saturation Measurements). The test regime for the combined dry and wet compaction measurements at room temperature, proposed as "in-situ saturation" measurements, consisted of loading-relaxation-saturation-relaxation-unloading stages, as displayed in Fig. 4.

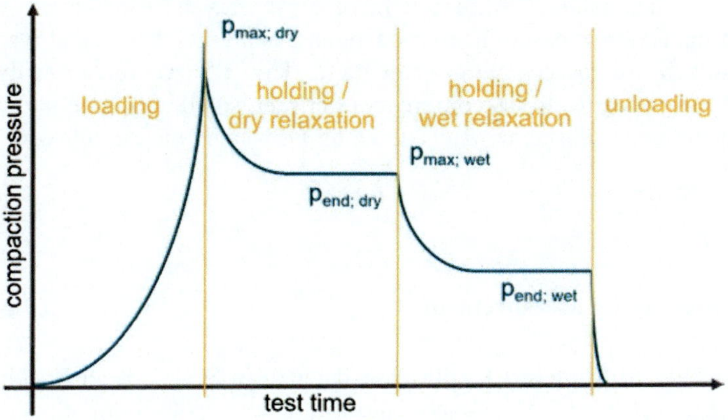

Fig. 4 Schematic of a compressibility measurement with an initial dry compaction and a secondary "in-situ" saturation. An optional heating phase for measurements above room temperature starts before the actual test time

During the loading phase, the specimen was compressed, according to [9] for better comparability, with a closing speed of 1 mm/s, to a calculated FVF of 0.54, resulting in the compressed specimen thickness of 3.000 mm. After reaching the set specimen thickness a relaxation period of 45 min follows, during which the crosshead position is held constant and the pressure decreases. After the dry relaxation period a valve was opened and a silicone oil started to saturate the specimen, beginning from the central injection point with a pressure of 0.2 bar. After an overall test time of 2.5 h, the unloading stage started, moving apart the test areas with an opening speed of 1 mm/s.

Binder Analysis. To better understand the results of the compression test at elevated temperatures further analysis of the raw binder as supplied by the manufacturer were conducted. These tests were aiming at quantifying the temperature level, at which the powder binder material changes its state from solid to viscous.

TGA/DSC-measurements. Based on the results of the compression tests, differential scanning calorimetry (DSC) tests and thermogravimetric analysis (TGA) of the binder were conducted. A Mettler Toledo DSC/TGA was used to get an overview of the binder behaviour across a wide temperature and time range, as well as two different test methods. A first heating cycle was used the get an impression of the stage of the raw binder, remove a material history and transform the binder from a powder to a quasi-monolithic form. Afterwards, a second heating cycle was able to measure the temperature characteristics of the binder without any previous influences.

Rheology Measurements. Rheological measurements were conducted on an Anton Paar MCR501 rotational rheometer at temperatures starting from 70 °C upwards in 5 K steps. However, below 95 °C the binder was completely solid and no measurement was possible. Between 95 and 100 °C the powder was still partly solid, therefore the measurement could not be evaluated. Since the data sheet presents a viscosity curve of the binder beginning at 100 °C, no further measurements were conducted.

Microscopy Measurements. Since the rheological measurements were not in line with the compaction measurements and binder analysis, the binder was analysed under an optical microscope with a hot stage. This allowed for an analysis of the binder particles, specifically the differences between small and large particles, under temperature influence during dynamic and long-term static measurements.

3 Results

3.1 Compressibility Measurements

Compressibility of bindered textiles. As detailed in Sect. 2.3, ten specimens were compressed to a FVF of 0.58 at each temperature level from room temperature up to 140 °C. The results of these measurements are displayed in Fig. 5.

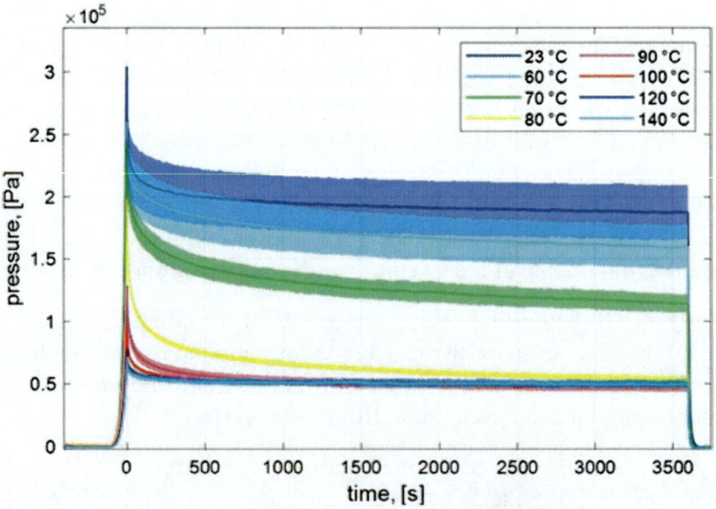

Fig. 5 Experimental data for dry compaction tests with bindered textile stacks at room and elevated temperature. Arithmetic mean of ten experiments (solid lines) with the corresponding standard deviation as envelope. The initial low compaction pressure represents the heating stage. Maximum compaction pressure was chosen as synchronization point for all experiments, after which the 1-h relaxation stage starts

With increasing temperature, the maximum compaction pressure (p_{max}) drops significantly. Above 100 °C, which corresponds to the specified binder processing temperature, the differences in p_{max} are small. In addition, beginning with a test temperature of 90 °C, constant pressure level is reached after about 1200 s of relaxation time. This remaining pressure after relaxation (p_{end}) is reached for 80 °C as well, but at a much later point during the relaxation stage. For lower test temperature levels, no constant pressure level was reached during the test time. This result allows for novel scientific insight, as up to now research focused on p_{max} only or just the first few minutes of relaxation.

Normalizing the pressure data with respect to p_{max} allows for the comparison of the speed of pressure decrease as well as the relative pressure drop during relaxation (Fig. 6). For tests between 70 and 100 °C, the relative pressure drop is significantly stronger compared to the remaining temperature levels, with a maximum of pressure relaxation of more than 70% found at 80 °C. The figure confirms that no constant pressure level can be reached within the test time of one hour for any temperature level, except for 140 °C.

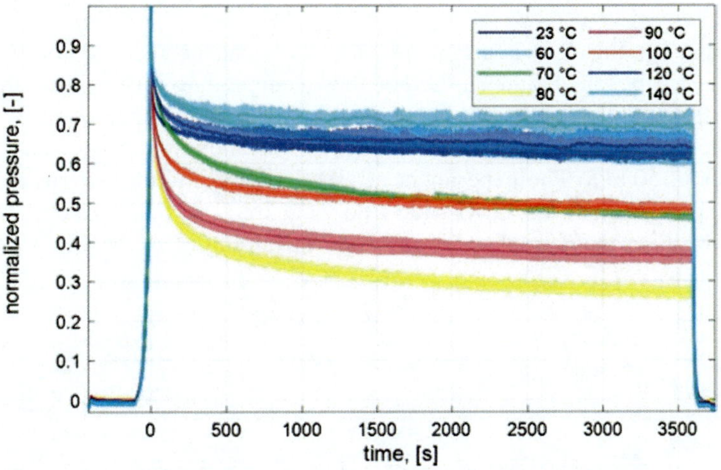

Fig. 6 Compaction results of dry compaction tests at room and elevated temperature levels normalized to p_{max}

As shown in Fig. 5 and 6 for most of the experiments, no steady state (constant pressure) was reached for the compaction pressure. Therefore, additional experiments with long-term relaxation periods of twelve hours were conducted at all temperature levels. The resulting pressure data are shown in Fig. 7.

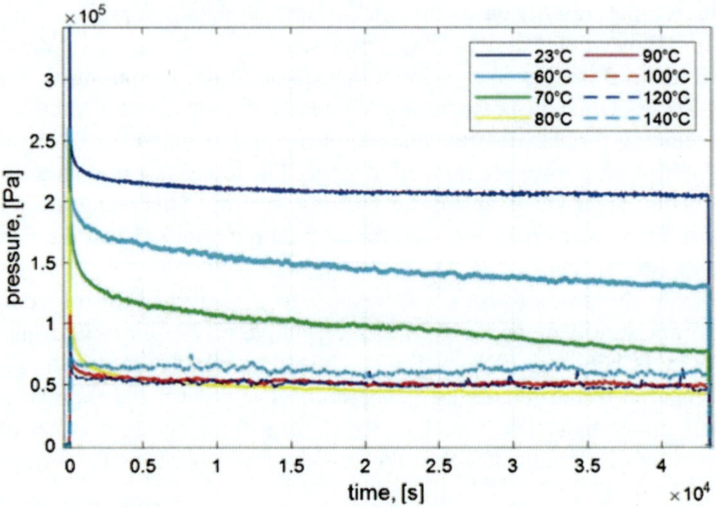

Fig. 7 Results of long-term dry compaction tests on bindered textiles at elevated temperatures

Compared to Fig. 5, a constant pressure level can be reached for all temperature levels, except for 60 and 70 °C. For these temperature levels, even after relaxation period of 12 h, a significant decrease of compaction pressure can be observed. Measurements for temperature levels of 80 °C and above reach a constant pressure level of about 50 kPa. The experiment at 70 °C shows a trend towards this level of 50 kPa. The experiment at 60 °C shows no clear trend to reach a constant pressure level. Fig. 8 shows the normalized pressure data from Fig. 7.

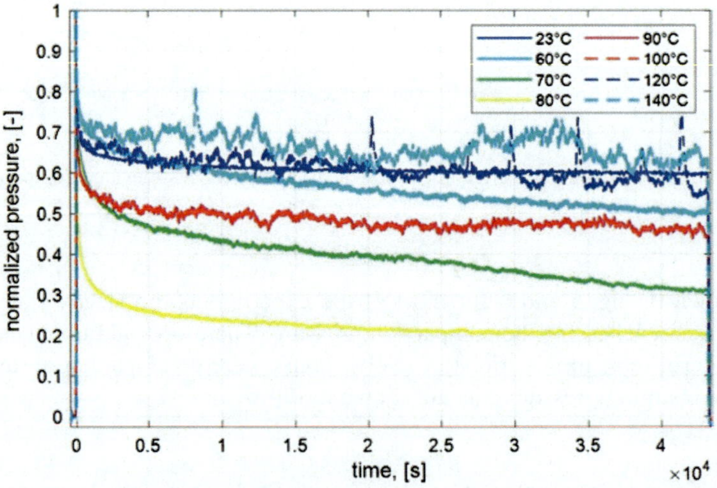

Fig. 8 Compaction results of dry, long term compaction tests at room and elevated temperature levels normalized to p_{max}

This figure shows more clearly that the extension of the relaxation period led to a constant pressure level for 80 °C after about 8 h. This relaxation to the 50 kPa level means a relative pressure drop (p_{max} - p_{end}) of more than 80%. An even bigger relative pressure drop can be expected for 70 °C The plots reveal unsteady characteristics in the data of the experiments run at 120 and 140 °C. This is probably due to corrections made by the UTM controller to compensate for changes in room temperature, caused by a long measurement period and measurements taken during a particularly hot summer period.

In-situ Saturation Measurements. In addition to the dry compaction set-up, this version of the test-rig was specifically designed to allow for a combined measurement of dry and wet compressibility behaviour. According to the test method, detailed in Sect. 2.3, compaction tests with in-situ saturation were conducted. The results of five tests at room temperature are displayed in Fig. 9.

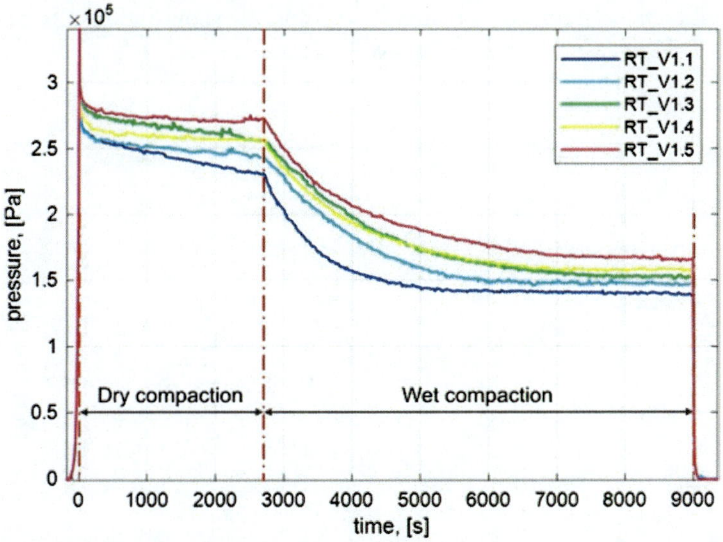

Fig. 9 Pressure relaxation behaviour of a bindered textile at room temperature in dry and wet conditions as a proof of concept for an "in-situ" saturation test method

Subdivided into a dry and a wet compaction stage, all five specimens show a relaxation behaviour after p_{max}, with one specimen reaching a constant pressure level after 2700 s, while the other specimen still show a significant pressure drop. This deviation can be attributed the common changes in the textile as well as the high number of layers, which allow for a high grade of fibre reorientation. At 2700 s the wet compaction stage begins and a valve is opened, allowing the measurement fluid to gradually saturate the specimen from the central injection point up- and outwards. As expected, a second stage of relaxation takes place with increasing levels of saturation across the specimen, allowing for further fibre reorientation, due to a lubrication effect of the measurement fluid. Overall, the five specimens show comparable wet relaxation behaviours, with small differences in the rate of the pressure drop, which can be explained by variations in stack permeability, resulting in different speeds of fluid propagation. After an overall test time of 2.5 h all specimen reached constant pressure levels.

3.2 Binder Analysis

Comparing the results of the compressibility tests with the technical data sheets of binder and textile, respectively, there is no clear agreement on the preferable processing temperature of the bindered textile. For this reason, the binder was tested in more detail to gain additional information that could explain the unexpected results of the compressibility measurements. Figures 10, 11 and 12 show the results of the combined TGA/DSC-measurements of the raw binder.

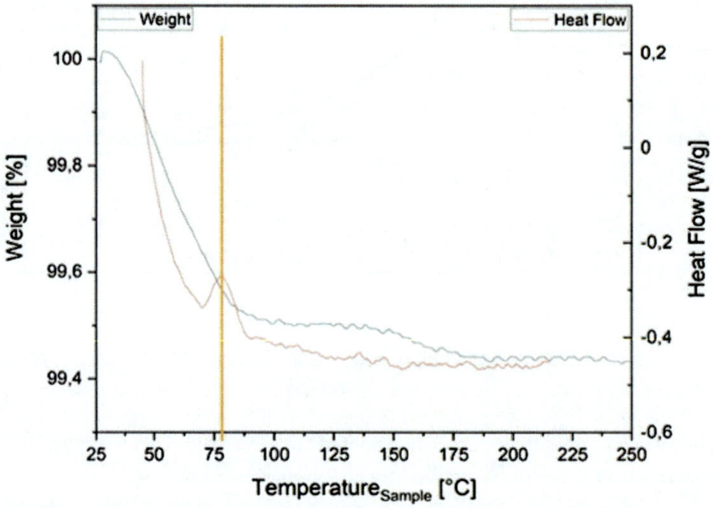

Fig. 10 TGA-measurement of the raw binder from 25 to 250 °C, showing an increased heat flow between 70 and 80 °C with the peak around 78 °C, without significant weight loss, indicating a change in the polymer structure. The measured weight loss of less than 0.6% can be attributed to evaporating air moisture, since the weight loss stabilizes at 100 °C

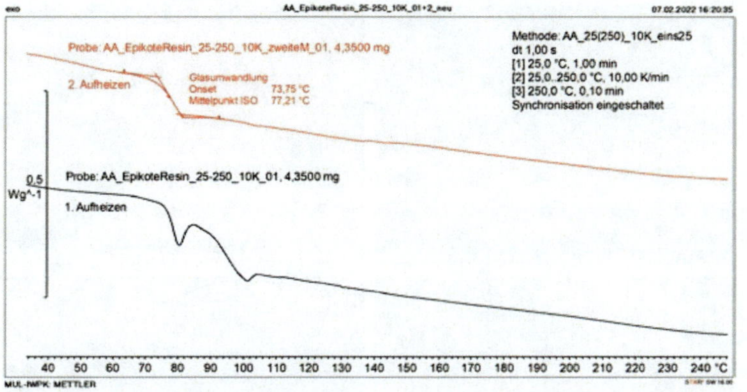

Fig. 11 DSC-measurement of the first (below) and second (above) heating cycle of the raw binder at a heating rate of 10 K/min from 25 to 250 °C, showing a T_G around 77 °C. A second dip in the first heating cycle at 100 °C cannot be found in the second cycle, but confirm the hypothesis of trapped air moisture in the raw binder

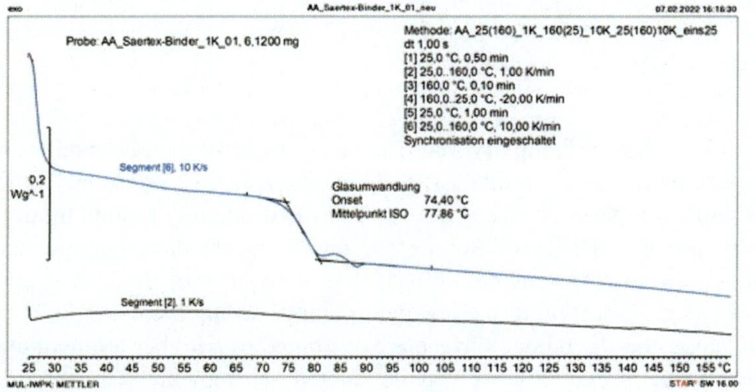

Fig. 12 DSC-measurement of the binder at a 1 K/min heating rate showing a T_G at 77.86 °C with an onset at 74.4 °C, matching the increased heat flow of the TGA-measurements

The results of the TGA/DSC-measurements indicate a change of state of the binder material around 80 °C, giving a first explanation of the relaxation behaviour 20 °C below the specified processing temperature of 100 °C. However, this does not explain the identical, but slower pressure relaxation at 60 and 70 °C. At these temperature levels, no change in material behaviour has been detected. For this reason, the binder was further analyzed through optical microscopy on a heated stage. Two different approaches were tested:(a) a dynamic measurement from room temperature up to 140 °C and (b) two long term measurements of one hour at 60 and 75 °C. Figure 13 shows a reference picture of the binder at room temperature and three elevated temperature levels.

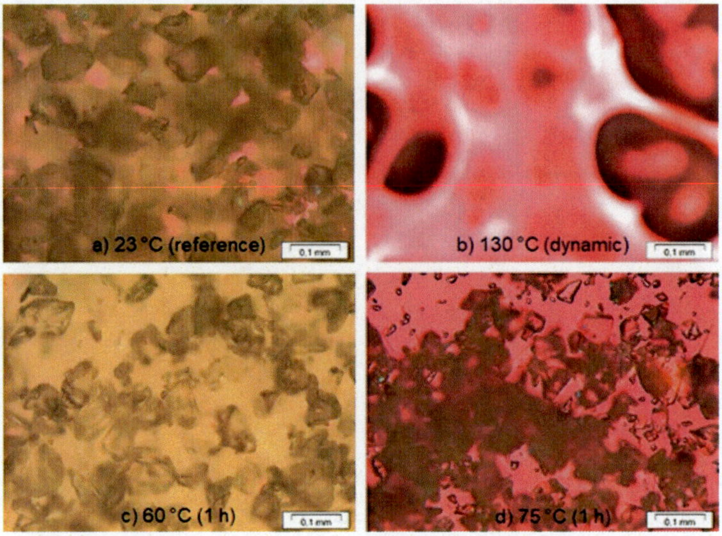

Fig. 13 Microscopic images of the binder at different temperature levels: Picture **a**) shows the raw binder at room temperature; picture **b**) shows the same area as in a) after a dynamic temperature sweep at 130 °C. The bottom pictures show the binder at temperature levels of 60 °C (**c**) and 75 °C (**d**) after one hour

Figure 13(a) shows exemplary microscopic sections of the raw binder with a wide distribution of particles sizes and mostly sharp edges. Figure 13(b) shows the situation after a dynamic temperature sweep at the point of completely molten binder particles, 130 °C. Compared to the TAG/DSC-measurements and the data sheet this temperature level is much too high. It was determined that the energy of the heat stage plate can only be transferred poorly into the powdered binder during the dynamic measurement and thus changes on the binder surface are attributed to a higher temperature. For this reason, two further measurements were conducted, during which the temperature level was held constant for one hour, allowing the binder to reach heat stage temperature. The temperature levels of 60 and 75 °C were selected, to get more information on the binder behaviour below T_G and T_M. It was hypothesized that smaller particles or the surface of bigger particles could already have crossed T_G and induce a partial lubrication effect, while most of the binder is still below T_G and does not contribute. However, at 60 °C (Fig. 13c) the particles appear similar to the reference and at 75 °C (Fig. 13d), even with a polarising filter to enhance the particle contours, no evidence, like rounded edges, can be found.

4 Discussion and Conclusion

4.1 Compaction Measurements

This paper presented a novel test-rig for compressibility measurements at elevated temperatures. Furthermore, this test-rig stands out from the current state-of-the-art, by offering an in-situ saturation method at room and elevated temperatures, greatly increasing the testing capabilities, while at the same time reducing testing effort.

The dry compaction measurements at room and elevated temperatures show a strong temperature dependence of the relaxation behaviour more than 30 °C below the specified processing temperature. In addition, a relaxation behaviour on long time scales was found, challenging the state-of-the-art approach on determining textile relaxation behaviour based on measurements on a time scale of minutes.

Further studies will (i) focus on confirming the binder influence on the relaxation behaviour through test with an unbindered NCF material (ii) expand the data by additional tests between room temperature and 60 °C to find an onset temperature for the lubrication effect. Moreover, creep tests will be conducted to determine, whether this behaviour not only occurs for relaxation, but also retardations measurements.

Finally, permeability measurements at the mentioned temperature levels will be conducted to determine, if the changes in relaxation behaviour, induced by an applied binder, also affect permeability at low temperature levels.

4.2 Binder Analysis

Comparing the results of the different analysis methods, only an inconsistent conclusion can be drawn at this stage. On one hand, the TGA/DSC-measurements confirm that above the specified processing temperature the binder is above T_G and T_M, thus existing in a state in which it can act as a lubricant between the fibres, and with that greatly reduces maximum compaction pressure. This behaviour was measured in the compaction tests above 100 °C. On the other hand, TGA/DSC-measurements, as well as the viscosity measurements and the microscopical observations at elevated temperatures, cannot explain the almost identical relaxation behaviour, in all but the time needed to reach the same level of relaxation, 30 °C below the measured binder T_G.

As a result of this, further analysis of the binder must be conducted, to explain the binder behaviour at low temperature levels.

Acknowledgements. The presented test-rig and the research work received funding through the project Evolution#4 from the Austrian research program "Take Off", which is a Research, Technology and Innovation Funding Program of the Austrian Federal Ministry for Climate Action, Environment, Energy, Mobility, Innovation and Technology, managed by the Austrian Research Promotion Agency (FFG), which is greatly appreciated.

References

1. Sousa, P., Lomov, S.V., Ivens, J.: Methodology of dry and wet compressibility measurement. Composites Part A: Applied Science and Manufacturing (2020). https://doi.org/10.1016/j.compositesa.2019.105672
2. Robitaille, F., Gauvin, R.: Compaction of textile reinforcements for composites manufacturing. I: Review of experimental results. Polym. Compos. (1998). https://doi.org/10.1002/pc.10091
3. Knight, J.C., Backes, D., Jayaraman, K.: Consolidation and relaxation behavior of continuous strand random glass mats with thermoplastic binders. Polym. Compos. (1996). https://doi.org/10.1002/pc.10633
4. Grieser, T., Breuer, U.P.: Textiles Formgebungsverhalten beim kontinuierlichen Preforming. Dissertation, Institut für Verbundwerkstoffe; Technische Universität Kaiserslautern
5. Li, L., Zhao, Y., Yang, J., Zhang, J., Duan, Y.: An experimental investigation of compaction behavior of carbon non-crimp fabrics for liquid composite molding. J Mater Sci (2015). https://doi.org/10.1007/s10853-015-8860-0
6. Robitaille, F., Gauvin, R.: Compaction of textile reinforcements for composites manufacturing. II: Compaction and relaxation of dry and H2O-saturated woven reinforcements. Polym. Compos. (1998). https://doi.org/10.1002/pc.10128
7. Dickert, M., Ziegmann, G.: Influence of binder on the mechanical properties and the permeability of a non-crimp carbon fabric preform. In: Proceedings ECCM15 - 15Th European Conference On Composite Materials
8. Wu, W., Jiang, B., Xie, L., Klunker, F., Aranda, S., Ziegmann, G.: Effect of Compaction and Preforming Parameters on the Compaction Behavior of Bindered Textile Preforms for Automated Composite Manufacturing. Appl Compos Mater (2013). https://doi.org/10.1007/s10443-012-9308-1
9. Yong, A., Aktas, A., May, D., Endruweit, A., Lomov, S.V., Advani, S., Hubert, P., Abaimov, S.G., Abliz, D., Akhatov, I., Ali, M.A., Allaoui, S., Allen, T., Berg, D.C., Bickerton, S., Caglar, B., Causse, P., Chiminelli, A., Comas-Cardona, S., Danzi, M., Dittmann, J., Dransfeld, C., Ermanni, P., Fauster, E., George, A., Gillibert, J., Govignon, Q., Graupner, R., Grishaev, V., Guilloux, A., Kabachi, M.A., Keller, A., Kind, K., Large, D., Laspalas, M., Lebedev, O.V., Lizaranzu, M., Long, A.C., López, C., Masania, K., v. Michaud, Middendorf, P., Mitschang, P., van Oosterom, S., Schubnel, R., Sharp, N., Sousa, P., Trochu, F., Umer, R., Valette, J., Wang, J.H.: Experimental characterisation of textile compaction response: A benchmark exercise. Composites Part A: Applied Science and Manufacturing (2021). https://doi.org/10.1016/j.compositesa.2021.106480
10. Hexcel: X-C-PB-555g/m^2-1270mm: Technical Data Sheet (2021)
11. Teijin Carbon Europe: HTS40 F13 12K line 2: Technical Data Sheet (2021)
12. Hexcel: 1202 TF970; Technical Data Sheet (2020)

Determination of the Bending Stiffness of Spread Carbon Fibre Tows Applied with Reactive Binder

Michael Liebl(✉), Mathias Engelfried, Stefan Carosella, and Peter Middendorf

Institute of Aircraft Design, University of Stuttgart, Pfaffenwaldring 31, 70569 Stuttgart, Germany
liebl@ifb.uni-stuttgart.de, engelfried@ifb.uni-stuttgart.de, stefan.carosella@ifb.uni-stuttgart.de, peter.middendorf@ifb.uni-stuttgart.de

Abstract. This paper deals with spread carbon fibre tows fixed by epoxy-based reactive binder that can be used in two-dimensional dry fibre placement (DFP) layups, which show freely selectable fibre orientation of each tow and layer structure. In a second step, the DFP preforms can be draped into a three-dimensional preform, here the bending stiffness has a significant influence on the forming behaviour. To build a mesoscopic draping simulation of the DFP preforms, detailed knowledge of the bending stiffness is necessary. A special test set-up for the bending stiffness characterization of bindered spread tows and experimental results are presented. As the bending behaviour is dominated by the epoxy-based thermoplastic binder, the investigations focus on the temperature dependent bending characteristics. Above the ideal forming temperature of the bindered spread tows, a decrease of the bending stiffness is visible. Additionally, the orientation of the one-side bindered tows is of significance for the bending stiffness. Beside to the experiments, a mesoscopic numerical model of the bending characterisation is set up. This model is used to validate the chosen material definition to correctly represent the bending stiffness of the bindered spread carbon tows in a later draping simulation.

Keywords: Fibre Placement · Draping · Preform · Composite · Simulation

1 Introduction

A common way to manufacture flat and moderately curved composite parts is done by forming two-dimensional semi-finished products into the final three-dimensional form, regardless of the use of prepreg or bindered dry material. When using dry material several options of textile products are available. One

J. Rieser et al. (Eds.): *Proceedings of the Munich Symposium on Lightweight Design 2021*, Proceedings, pp. 59–67, 2023. https://doi.org/10.1007/978-3-662-65216-9_6

opportunity is the use of spread carbon fibre tows, which are used in a dry fibre placement (DFP) process to generate two-dimensional layups. Benefits of this manufacturing process are high reproducibility as well as flexibility in fibre orientation while causing low waste by near-net-shaping [1]. During the following forming of these DFP preforms the individual spread carbon tows are bent regarding the surface topology of the mould. The aim of this research is to determine the bending behaviour of the spread carbon tows with respect to different possible forming temperatures. To establish a process simulation, a numerical model of the bending test is developed. The experimental results are used to determine the model's parameters to optimally describe the material behaviour during bending.

2 Material and Methods

The examined material, M&A Dieterle FixedTow, is a spread carbon tow with an average width of 20 ± 0.5 mm [2]. The used basic tow is SIGRAFIL®C T24-5.0/270-E100, a 24k roving from SGL Carbon, which has been bindered on one side for fixation and improved processability in later steps like DFP. As binder Hexion Epikote™ Resin TRAC 06720 was used during the production of the FixedTows with an average binder amount of 8 wt.% or equivalent 9 g/m^2. This reactive epoxy-based binder starts melting in the temperature range of 80–90 °C and behaves thermoplastic. Above a temperature of 100 °C cross-linking starts. The binder is well suited for further use in the forming process [3].

In order to be able to determine the bending stiffness of the spread carbon tows as realistically as possible, the FixedTows will be laid down individually with the Crosslayer process and a length of 30 cm before the actual characterisation. As the FixedTow will be unrolled, briefly heated and pressed to the substrate in the Crosslayer during the DFP, this circumstance has to be considered when determining the bending stiffness of the FixedTows for the following forming process.

2.1 Cantilever

A widely used method to determine the bending stiffness of textiles is the simplified cantilever test, which is defined in DIN 53362:2003-10 or with minor varieties in ASTM D1388-07. The following test method is referring to DIN 53362:2003-10, but with slightly deviations for example in the given specimen width of 20 mm instead of minimum 25 mm [4].

In Fig. 1 the setup of the cantilever test is schematically shown, where the test specimen is moved to the right over the edge. As soon as the specimen touches the 41.5° slope the test is stopped and the overhang length $l_{ü}$ is documented. In

DIN 53362 [4] the bending stiffness B is described as a value of resistance of a specimen, which it opposes to the attempt of bending due to its own weight. It is calculated with F_1 as fabric weight per length in Eq. (1)

$$B = F_1 \cdot \left(\frac{l_{ü}}{2}\right)^3 = m_f \cdot b \cdot \left(\frac{l_{ü}}{2}\right)^3 \cdot g \tag{1}$$

With the given areal mass $m_f = 89\,\frac{g}{m^2}$ of the FixedTow, the width b of the specimen and the gravitational acceleration g, the bending stiffness can be determined [3].

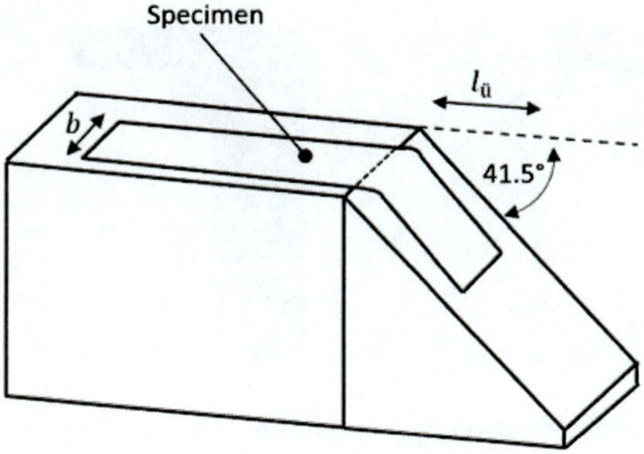

Fig. 1. Cantilever setup referring to [4]

2.2 Optical Test Arrangement

As the bending stiffness characterisation is realised with respect to the temperature influence, it is necessary to ensure a homogeneous heat distribution during the measurement. Therefore, the tests are performed in a *Memmert UF260plus* convection oven, which provides a constant environmental temperature. The automatized cantilever setup (cf. Fig. 3) is unsuitable for the use in the oven, because of the not heat-resistant electronic system and its overall dimensions. Therefore a alternative approach of determining the bending stiffness of the FixedTow is established.

Basics. In Fig. 2 the test arrangement inside the oven is shown, where a FixedTow specimen is clamped on one side and free-hanging due to its own weight. This equals a cantilever beam, which displacement curve can be described by Eq. (2) with respect to Bernoulli's assumptions [5]. Here w describes the displacement of the beam, which has a bending stiffness of EI, under the load of the moment M.

$$w'' = -\frac{M}{EI} \tag{2}$$

With the given boundary conditions for the fixed cantilever beam and a constant line load q_0 due to the own weight of the beam, the displacement of the beam can be described as shown in Eq. (3). Parameter l equals the length of the undeformed beam [5]. In case of the FixedTows, q_0 is determined by Eq. (4).

$$w(x) = \frac{q_0 l^4}{24EI} \cdot \left[\left(\frac{x}{l}\right)^4 - 4\left(\frac{x}{l}\right)^3 + 6\left(\frac{x}{l}\right)^2 \right] \tag{3}$$

$$q_0 = m_f \cdot b \cdot g \tag{4}$$

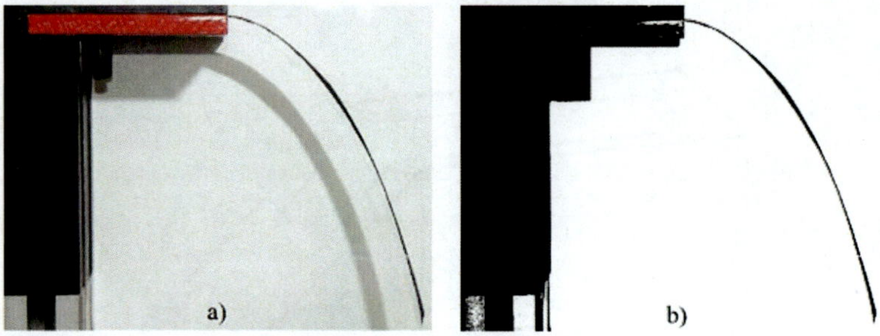

Fig. 2. Images inside the oven: a) original, b) black and white transformation

Evaluation. To analyse the bending stiffness of the FixedTows, a detailed image (cf. Fig. 2) is taken inside the oven. Thereby, a white background and an illumination to avoid shadows as good as possible are necessary. The clamping device has a defined red area with a length of 10 cm as a scaling factor for the following optical analysis. The images are done with a *Canon EOS 600D* with a *Canon EFS 18–55* mm objective.

First the picture is transformed into a black and white image as shown in Fig. 2. Based on this image the displacement curve $w(x)$ can be determined (cf. Fig. 6) and used for the evaluation of $EI(x)$ using Eq. (3). Finally, the specimen's bending stiffness EI results as the average value of $EI(x)$.

3 Bending Stiffness

3.1 Cantilever Reference

For a first estimation of the spread tow's bending stiffness, classical cantilever tests are carried out referring to DIN 53362:2003-10 at room temperature. For this purpose, the existing setup as shown in Fig. 3 is used. The movement of the specimen is controlled via a electronic system to ensure a constant velocity. Five specimens are tested with an average bending stiffness of $B = 31.85 \pm 1.09$ Nmm^2.

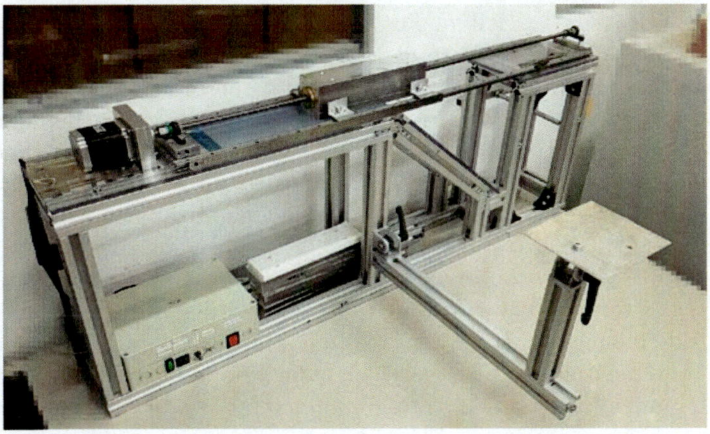

Fig. 3. Cantilever setup at the Institute of Aircraft Design

3.2 Optical Test Arrangement

Since the determination of the bending stiffness is based on the displacement curve, it is necessary to validate the use of the optical test arrangement. During the cantilever tests at room temperature, pictures of the bent FixedTow are taken, which are then analysed with the optical method and the displacement curve is determined. Here, the average bending stiffness of five specimens is $EI = 29.84 \pm 1.52$ Nmm^2, which shown no significant difference to the results of the cantilever test.

The principle of the optical test arrangement requires to primarily define the length of the unbent specimen, which equals parameter l from Eq. (3). This length is chosen to $l = 200$ mm for all tests. To check again the validity of the optical evaluation method, the bending stiffness of the FixedTow is evaluated at ambient temperature inside the oven. The determined bending stiffness of $EI = 30.74 \pm 4.04$ Nmm^2 shows sufficient conformity to the comparable evaluation referring to DIN 53362 in Sect. 3.1.

3.3 Temperature Variation

As the Epikote^{TM} 06720 binder shows thermoplastic forming behaviour at temperatures between 80 °C and 90 °C, it is designated to drape the FixedTow textiles in this range of temperature. First draping tests with a punch-and-die geometry with the mold heated to 80 °C indicated less draping defects like gaps and wrinkles in comparison with forming at ambient temperature. Therefore, the focus in determining the bending stiffness of the FixedTows is on temperatures around the proposed forming temperature.

Figure 4 displays the determination of the FixedTow bending stiffness inside the convection oven at temperatures from 75 °C to 105 °C by using the optical test arrangement. In general, a slight trend of decreasing bending stiffness with increasing temperature is visible. The orientation of the FixedTow has a higher influence on the bending stiffness at lower forming temperatures at 75 °C and 80 °C. Here, the bending stiffness is increased if the bindered side of the FixedTow is placed downward. With rising temperature above the binder's melting temperature the orientation of the FixedTow has no significant effect on the bending stiffness.

This behaviour may originate from the binder's structure, which is partially applied over the surface of the spread carbon tow. In the case of low temperature as well as binder on top of the bent FixedTow the particles can easily drift apart under the tensile load, resulting in a low bending stiffness. Is the bindered FixedTow side facing downward a pressure load is applied and the binder particles inhibit the bending of the tow, which leads to a higher bending stiffness. As soon as the binder's reversible forming temperature is reached, the binder starts melting and enabling a sliding of the fibres.

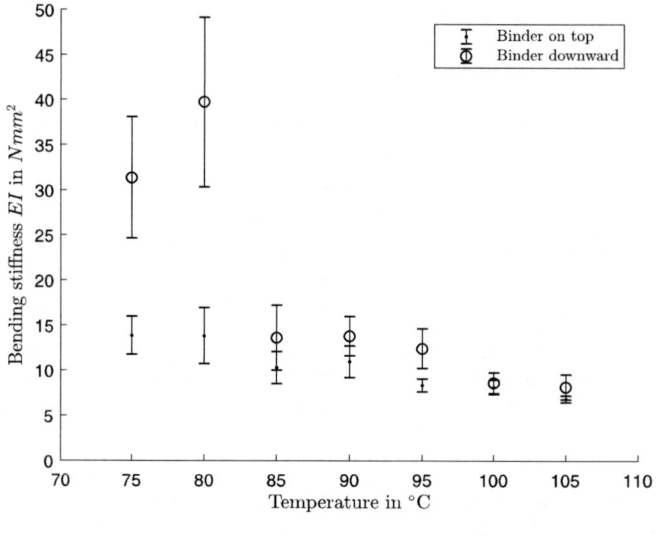

Fig. 4. FixedTow bending stiffness in the range of the forming temperature

4 Numerical Approach

For a later numerical process simulation of the textile draping process the cantilever test is modelled in order to validate the chosen material model. The software tools from Dynamore GmbH are used for modelling and computation.

4.1 Model and Material

The explicit simulation model describes the simplified cantilever test inside the oven, where the specimen is fixed at one side and free-hanging under gravitational load. Shell elements are used to describe the specimen and plate of the test arrangement.

LS-Dyna material MAT_FABRIC (Material Type 34) is used for this simulation. This material model was developed for airbag materials and is equipped with a membrane element formulation which suits large deformation of fabrics [6]. The material model provides no bending resistance and thus the elements can fold undamped [7]. To consider the bending stiffness of the material, an additional card in the material deck is activated. This creates a virtual coating of the material with a defined offset from the membrane. During deforming of the elements the coating deforms along and thus adding a rotational resistance equivalent to the specimen's bending stiffness. Input parameters of the coating are the Young's Modulus, yield stress and layer thickness [6,7].

In this research, the coating thickness $TCOAT$ is chosen to 0.01 mm, which is equal to 10% of the thickness of the spread carbon tow specimen. Yield stress $SCOAT$ is predefined to 3.5 GPa. For the first estimation of the Young's Modulus $ECOAT$ the membrane and coating combination is assumed to be an ideal I-beam. With the geometric dimensions, width as well as thickness, of the specimen the area moment of inertia I is determined. With the given bending stiffness EI, the Young's Modulus is the quotient from EI and I. The experimental results show a nearly equal bending stiffness of the spread carbon tow at a temperature of 85 °C regardless the tow's orientation. In the later forming process this temperature will be chosen to achieve a satisfying quality of the textile. Therefore, the focus of the process simulation is upon this temperature. For the numerical approach the measured bending stiffness $EI = 10.3$ Nmm^2 at 85 °C of the spread carbon tow (Binder on top) is taken into account, because at this temperature the material behaviour is independent of the tow's orientation. Therefore, it is sufficient to model a bending test with the material's virtual coating considered as $ECOAT = 19$ GPa. Figure 5 shows the simulation model of the tow during the bending test.

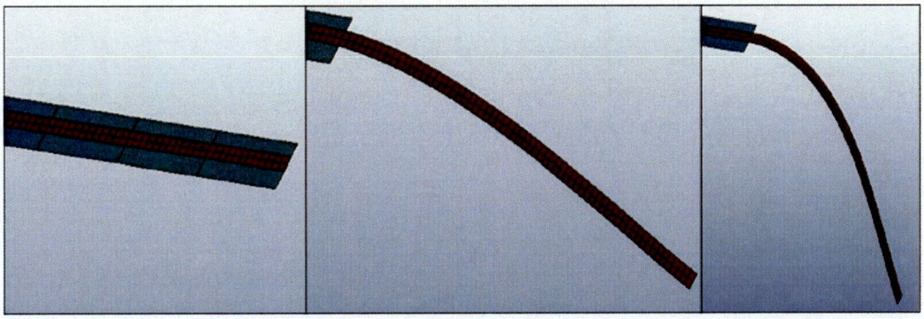

Fig. 5. Initial state, mid simulation and final bending state (f.l.t.r)

4.2 Results

A variation of the parameter $SCOAT$ has no visible influence on the bending behaviour of the model and is therefore set constant to 3.5 GPa. In Fig. 6 the displacement $w(x)$ of the numerical model with $ECOAT = 19$ GPa is superimposed with the experimental result.

The first assumption of $ECOAT$ is clearly not optimal to model the bending stiffness of the carbon spread tow. Here, two main reasons are possible. First the determination of the bending stiffness EI with the optical test arrangement. Especially the analytical approach according to Eq. (3), uses influential assumptions according to Bernoulli. Second the transformation of the bending stiffness EI to the virtual coating's Young's Modulus $ECOAT$ of the MAT_FABRIC is a rough estimation. In the further proceeding the parameter $ECOAT$ is incrementally decreased to achieve a satisfying alignment of the numerical model to the expermental results. Finally, the parameter $ECOAT$ is set to 9 GPa as shown in Fig. 6.

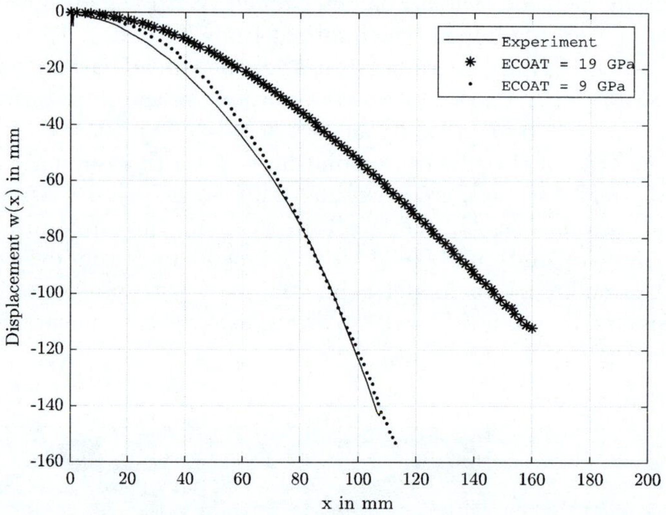

Fig. 6. Displacement curves at 85 °C with binder on top

5 Conclusion

To determine the temperature dependent bending stiffness of a spread carbon tow, an alternative method is presented. The spread tow specimen is hanging free with one side clamped. With the help of picture analysis, the displacement of the specimen over its length and its bending stiffness is evaluated. The one side bindered carbon spread tow shows a decreasing bending stiffness with rising temperatures. At temperatures of 75 °C as well as 80 °C a significant difference in bending stiffness with varying tow orientation is visible. If the bindered side of the tow is facing downward a much higher bending stiffness occurs compared to binder on top. At and above 85 °C, no clear distinction between the two tow sides can be made. Potentially, this behaviour has its origin in the structure of the binder. Under pressure load, with the bindered tow side downward, the binder inhibits the bending. Above the melting temperature of the epoxy based binder its interference with the bending stiffness is decreasing. The forming of the spread tow based textiles will be carried out at 85 °C.

A numerical process simulation of the textile forming will be established in the LS-Dyna environment, to identify the optimal process parameters. Therefore, a material model has to be chosen, which represents the behaviour of the carbon spread tow. With the experimentally determined bending stiffness, the material MAT_FABRIC respectively the parameter $ECOAT$ is calibrated. For the desired forming temperature of 85 °C a satisfying alignment of model and experiment can be achieved.

Acknowledgments. The authors thank the German Federal Ministry of Education and Research for making this research possible by funding the research project DIREKT (FKZ: 03INT710AA) and M&A Dieterle GmbH for the material supply.

References

1. Grisin, B., Carosella, S., Middendorf, P.: Dry Fibre Placement - Influence of Process Parameters on Mechanical Laminate Properties and Infusion Behaviour. 27. Stuttgarter Kunststoffkolloquium (2021)
2. UD FixedTow Production Line. www.ma-dieterle.de/images/dieterle/verbund/Broschueren/Broschuere19_UD_FixedTow.pdf (visit: 05.05.2021)
3. UD-fixedTow CF 24k. www.ma-dieterle.de/images/dieterle/verbund/BroschuerenfixedTow_SGL_24k_8Epikote06720_MADieterle.pdf (visit: 05.05.2021)
4. DIN Deutsches Institut für Normung e. V.: Bestimmung der Biegesteifigkeit - Verfahren nach Cantilever, DIN 53362:2003-10 (2003)
5. Gross, D. et al.: Technische Mechanik 2. 11. Auflage. pp. 87–173. Springer-Verlag, Berlin Heidelberg (2012). https://doi.org/10.1007/978-3-642-19984-4
6. Livermore Software Technology Corporation: LS-DYNA Keyword User's Manual - Volume II - Material Models - LS-DYNA R11. Livermore, California (2018)
7. Borrvall, T., Ehle, C., Stratton, T.: A Fabric Material Model with Stress Map Functionality in LS-DYNA. 10th European LS-DYNA Converence, Würzburg (2015)

Manufacturing Technologies for Box-Shaped Pressure Vessels with Inner Tension Struts

Christian Wrana[1], Konstantin Heidacher[1], Michael Ruf[2,3], Dominik Joop[1], and Alexander Horoschenkoff[1](✉)

[1] Laboratory of Polymer Technologies, Munich University of Applied Sciences, 80335 Munich, Germany
{wrana,konstantin.heidacher0, alexander.horoschenkoff}@hm.edu

[2] Chair of Carbon Composites, Technical University of Munich, 85748 Garching, Germany
Michael.G.Ruf@bmw.de

[3] Department of Hydrogen Supply, BMW AG, 80937 Munich, Germany

Abstract. Shared vehicles architectures for fuel cell and battery electric vehicles offer a high potential for cost reduction by enabling economies of scale in engineering and production. The efficient integration of hydrogen storages in flat box-shaped battery design spaces represents one of the essential basic requirements. As state-of-the-art cylindrical pressure vessels do not allow a high volumetric efficiency in the installation space, two concepts of box-shaped pressure vessels with tension struts are investigated with regard to manufacturability. The first concept focuses on the integration of aramid fibers in a carbon fiber tank by tufting. In a second concept 3D weaving is analyzed with regard to the construction of a pressure vessel with inner tension struts. For both tank designs manufacturing technologies are developed and the concepts are validated using prototypes. Considering technologies for series production of the textile sector possible paths for industrialization are identified.

Keywords: Pressure Vessel · Hydrogen · Aramid Fiber Strut · Tufting · 3D Weaving

1 Introduction

1.1 Motivation

With the Green Deal the European Union commits to a significant reduction of CO_2 emissions within the next decades to become the first climate-neutral continent of the world by 2050 [1]. This requires solutions for decarbonization across all sectors. In transportation, emission free solutions are needed for public and individual mobility. Currently car manufacturers mainly focus on battery electric vehicles (BEVs) as a

J. Rieser et al. (Eds.): *Proceedings of the Munich Symposium on Lightweight Design 2021*, Proceedings, pp. 68–82, 2023. https://doi.org/10.1007/978-3-662-65216-9_7

solution for zero emission vehicles [2]. But besides BEVs fuel cell electric vehicles (FCEVs) offer a high potential for an emission free mobility and additionally support sector coupling for the storage, transportation and use of renewable energies [3, 4]. Furthermore, FCEVs offer advantages like high vehicle ranges and short refueling times [5]. However, compared with BEVs, FCEVs are still a niche product due to a limited infrastructure of hydrogen filling stations and high vehicles prices [6]. To increase the attractiveness of fuel cell electric vehicles for customers, the costs must be decreased.

Shared vehicles architectures for BEVs and FCEVs enable economies of scale in engineering and production and allow a significant reduction of costs for both powertrain variants. This is only possible if both vehicle types share similar design spaces for their drivetrain components. In current BEVs the battery system is integrated in a flat box-shaped design space in the car underbody for reasons of safety and modularity (Fig. 1) [6]. In order to use the same vehicle architecture for FCEVs, the hydrogen storage must fit in a similar design space. To reach customers´ expectations according to vehicle ranges, an efficient integration of the hydrogen storage is a main requirement. Current storage technologies are based on cylindrical pressure vessels and do not allow an efficient exploitation of the design space. Therefore, new pressure vessel technologies with an adapted geometry are required to improve the volumetric efficiency of the hydrogen storage.

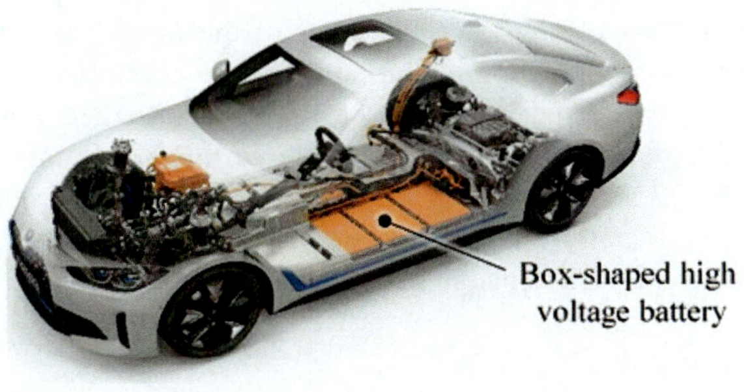

Fig. 1. Box-shaped design for the integration of a battery system in a battery electric vehicle, illustrated by the example of a *BMW i4* [7]

1.2 State-of-the-art

Today, the development of FCEVs is mainly based on vehicle architectures for combustion engine technologies. In this case, the fuel cell is integrated in the front carriage instead of an engine. The hydrogen gas is stored in two or three pressure vessels, which are mounted either lengthwise in the middle tunnel instead of the cardan shaft or transversely in front of and behind the rear axle instead of conventional fuel tanks (Fig. 2). In the automotive sector, a nominal working pressure of 70 MPa has become a standard for storing the gaseous hydrogen under high pressure. Therefore, usually type IV pressure vessels are used, which consist of a carbon fiber reinforced plastic (CFRP) tank wall to enable an efficient and load-appropriate design. On the inside of the tank wall a polymer liner is used for sealing the containment and restrict hydrogen permeation. To assure full safety during operation the hydrogen vessels are designed for a minimum burst pressure of 157 MPa. This corresponds to a safety factor of at least 2,25 regarding the nominal working pressure of 70 MPa [8].

Fig. 2. State-of-the-art fuel cell electric vehicle (FCEV) with 70 MPa tank technology, illustrated by the example of a *Toyota Mirai 2* [9]

1.3 Concept of a box-shaped pressure vessel with inner tension struts

A perfectly cuboid pressure vessel does not allow an efficient vessel design due to the internal pressure load. Due to the geometry stress peaks occur in the corners and edges of the pressure vessel as a result of the load, as well as strong deformations of large planar surfaces. Therefore, the edge areas of the investigated vessel design are rounded off and the large opposing planar surfaces are connected with tension struts. The strut loading is transferred by a loop design using a deflector support structure. Analytical and numerical investigations have shown how a design for such a vessel construction can be carried out [10, 11]. For the integration into a flat cuboid battery storage space, a significantly higher storage potential is possible for the box-shaped pressure vessel concept with tension struts than for conventional cylindrical vessel designs [12].

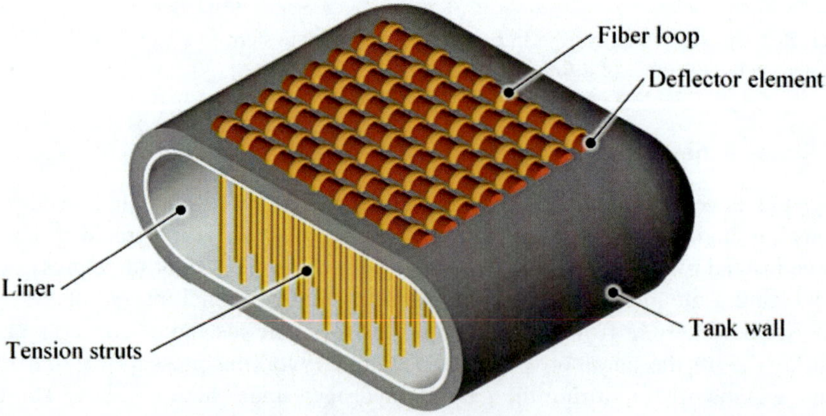

Fig. 3. Conceptual representation of a pressure vessel with tension struts in the interior

2 Manufacturing concepts

Starting from the conceptual design of a cuboid pressure vessel with inner tension struts, two manufacturing technologies are developed which lead to different design characteristics of the vessel.

2.1 Differential pressure vessel design

The first concept is a differential design in which the fabrication of the CFRP tank wall is decoupled from the insertion of the tension struts. The integration of the aramid fiber struts is done by tufting. An overview of the entire manufacturing process is shown in Fig. 4.

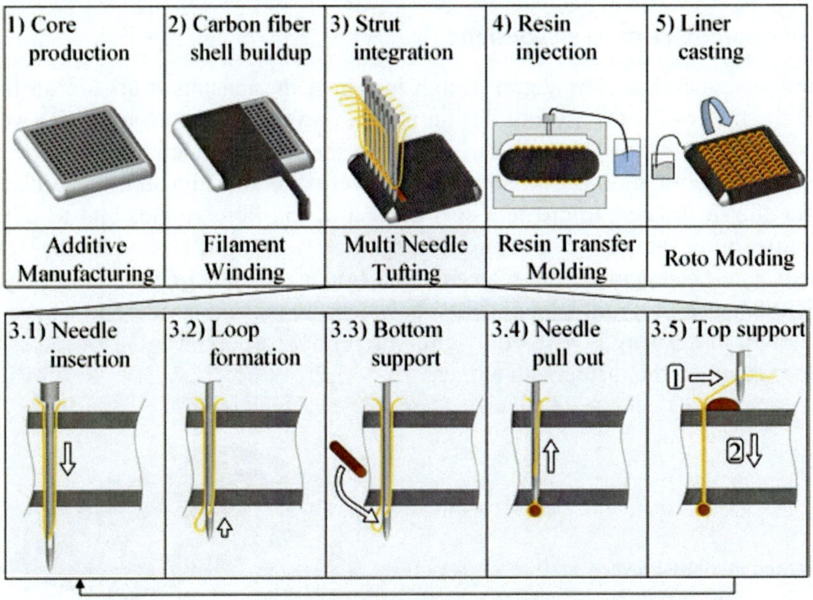

Fig. 4. Manufacturing concept of a pressure vessel by means of a tufting process as core technology for the insertion of tension struts

The manufacturing for the differential pressure vessel design is divided into five steps: In the first step a two-part core is manufactured in an additive manufacturing process (1). A grid structure on the upper and lower side serves to guide the needles during the tufting process. In the second step carbon fiber bandages [0° and 90°] are wounded around the core (2). These layers absorb the horizontal forces acting on the tank. The third step is the integration of the tension struts using a tufting process (3). Several needles are arranged parallel to each other with a distance of 4 mm. The first tuft stitch starts at one end of the planar surface. Two different loop shapes are formed by the tufting process: The needle feed in vertical direction creates a tufting stitch (step 3.1 to 3.4) and the feed in horizontal direction creates a sewing stitch (3.5). The first step of the tufting process is the penetration of the entire vessel by the needles in vertical direction (3.1). Afterwards, the needles move back slightly, and a loop tunnel is formed by the aramid fibers (3.2). A deflector element is inserted into the loop tunnel (3.3). Subsequently, the loop tunnel is closed and the needles move to the start position (3.4). The deflector element is pressed against the carbon tank wall. Finally, the needles are moved 4 mm in the horizontal direction (3.5). After the movement the deflector element is inserted and is integrated with the next sewing stitch (3.1). This process is repeated for the entire planar surface of the tank wall. The fourth step is the impregnation of the carbon fiber bandages. The impregnation is carried out in two steps (top and bottom) based on the RTM process (4). Finally, roto molding is used for the insertion of the liner (5).

2.2 Integral pressure vessel design

The second concept is an integral design in which the tension struts are an integral part of the tank wall. The overall design of this design orientates on the 3D-weaving process and is shown in Fig. 5. In this principle, the deflector elements are woven into the illustrated fibre strands and serve as the weft direction. Ideally, these deflectors have an elliptical shape to create a soft deposit of the fibre strands and to minimise the ondulation of the woven structure. The pressure loading is carried by the deflector elements and the support of the filler material (green) by the circumferential layers. A constant distance of 4 mm is maintained between the tension struts. The woven structure built up in this way is covered by circumferential carbon layers in x- and y- direction and sealed on the inside with a liner.

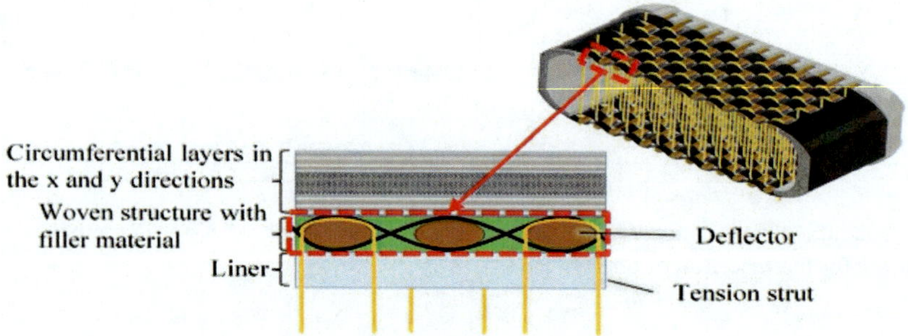

Fig. 5. Cross section and sectional view of the 3D-woven structure

Figure 6 shows the 3D-weaving process of the vessel. The two-part core is manufactured using an additive manufacturing process (1). One side element of the core can be removed so that the strut integration can be performed from this side. The insertion of the tension struts then takes place in five steps (2). By alternately moving the fibres, one half is moved to the upper position and the other half to the lower position, thus creating the characteristic shed opening (2.1). By alternately moving the tension struts, one half is moved to the upper position and the other half to the lower position, thus creating an X-structure (2.2). With the help of a pull-back-device, the tension struts are pulled into position (2.3). The deflectors can now be attached at the top and bottom through the shed opening (2.4). By removing the pull-back, the tension struts wrap around the deflectors and remain fixed in place (2.5). These five steps are repeated until the end of the core is reached and thus the inner structure of the vessel has been built up. After the finalization of the weaving process the last side element of the core can be attached. Further steps follow to complete the pressure vessel. The impregnation of the liner takes part (3). The liner is integrated in the first step with a sink moulding process, in which the upper and lower woven structure is sealed. For this purpose, the vessel is placed in a mould that is filled with the liner material. The open weave structure allows the liner material to penetrate the structure and form a uniform level layer inside the vessel. In the second step, a roto moulding process is used. In this process the vessel is rotated, which distributes the liner material inside the vessel and forms a round liner structure. Thus, with the help of the combination of both liner insertions, a defined liner structure is created inside the vessel. The vessel is finished by applying a fibre shell using a wet winding process, which is compressed with the help of a vacuum bag process.

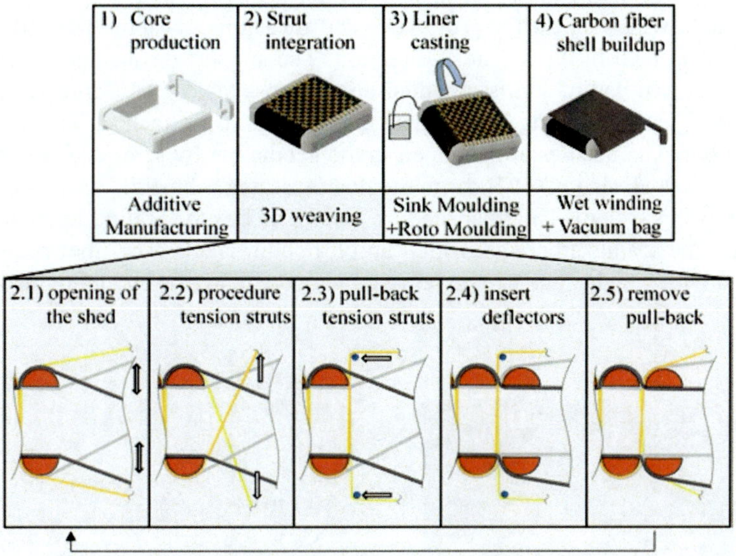

Fig. 6. Manufacturing concept of an integral pressure vessel design with 3D weaving as core technology for the insertion of tension struts

3 Manufacturing of Prototypes

At the Munich University of Applied Sciences, the first hydrogen pressure vessel prototypes are developed on a scale model. The basis of the scale model is a 3D printed PLA core and the boss of the vessel is manufactured from an aluminum tube that is bonded to the additively manufactured core. The carbon fiber type T700 from the manufacturer Toray is used for the winded layers. The stacking of the winded layers is [0,90]sym and creates a biaxial layer structure. A D2200 aramid fiber from the manufacturer Teijin is used for the tension struts. For better processability of the aramid fiber it is twisted with 250 rotations per meter. The deflectors are made of metal with a round shape. The liner insertion and the impregnation of the circumferential layers is done with an epoxy resin.

3.1 Tufting

The dimensions of the final prototypes are $91 \times 91 \times 31$ mm and the strut spacing of 4 mm results in a pattern of 16×16 tension struts. The tension struts are inserted using a developed tufting machine. The tufting machine consists of the following components (Fig. 7):

1. Spindle drive: The force- and time-controlled spindle drive of a tensile testing machine (Zwick 1465) is used to realize the needle bar feed. With this spindle drive a force and speed monitoring is possible.
2. Cross table with support for prototypes and measuring system: The cross table is used for the feed in the x- and y-direction. The support device for the prototypes consists of two halves and is mounted on the cross table. For a precise position of the core a glass measuring system with digital display is attached.
3. Needle bar: The needles are clamped in the needle bar by screws. The used needle type is a 1000H, size NM 250 from the manufacturer Schmetz.
4. Fiber feeding system: The fiber feeding system guides the aramid fibers to the needle bar. The aramid fibers are rolled up on bobbin creels. The fiber pre-tension is realized with weights and can be released with the help of a clamping mechanism.

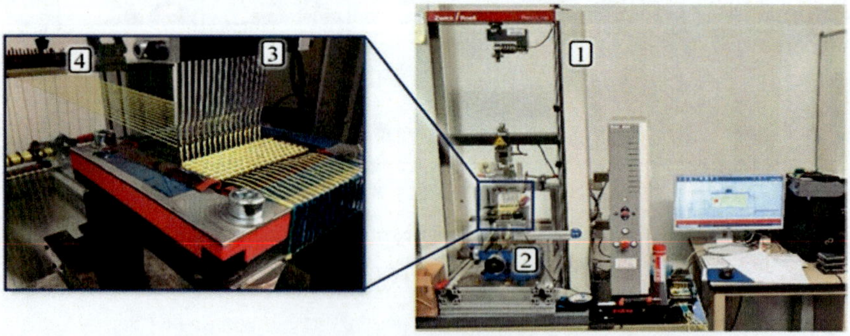

Fig. 7. Tufting machine for the insertion of the tension struts

With this machine, it is possible to manufacture tanks with internal tension struts. Figure 8 shows such a tank, which has been cut open on two sides for visualization.

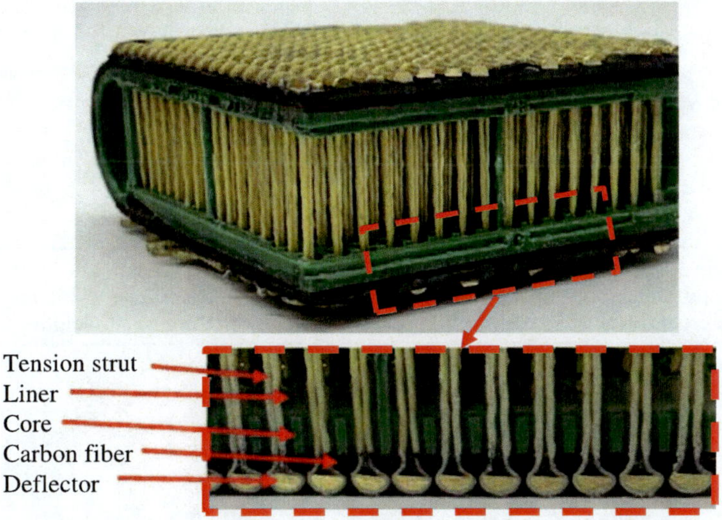

Fig. 8. Manufactured prototype of the tufting in sectional view

3.2 3D-weaving

The prototypes made with the 3D-weaving have a size of 100 mm × 100 mm and a height of 25 mm, resulting in a strut pattern of 12 × 17 struts. The tension struts are inserted using a developed weaving machine. This weaving machine is based on the principle of a loom and has the dimensions 1,5 m long, 1 m wide and 1 m high. The machine basically has the following four sections which are shown in Fig. 9:

- weaving place: including the core fixation
- strut insertion: the tension struts are guided through the machine and inserted into the vessel
- shafts: realisation of the characteristic shed opening
- fibre feeding system: supplies the machine with the fibres and puts them under tension

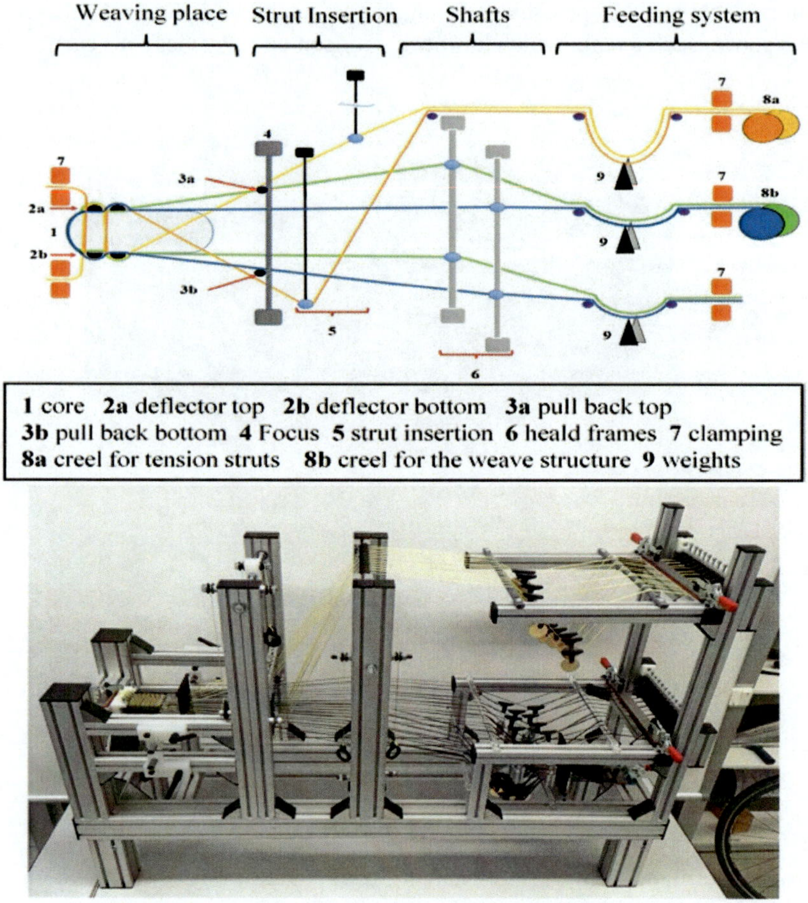

Fig. 9. Sketch and side view of the whole weaving machine

With the help of this machine, it is now possible to manufacture tanks that have internal tension struts. Figure 10 shows such a tank, which has been cut open on two sides for visualization.

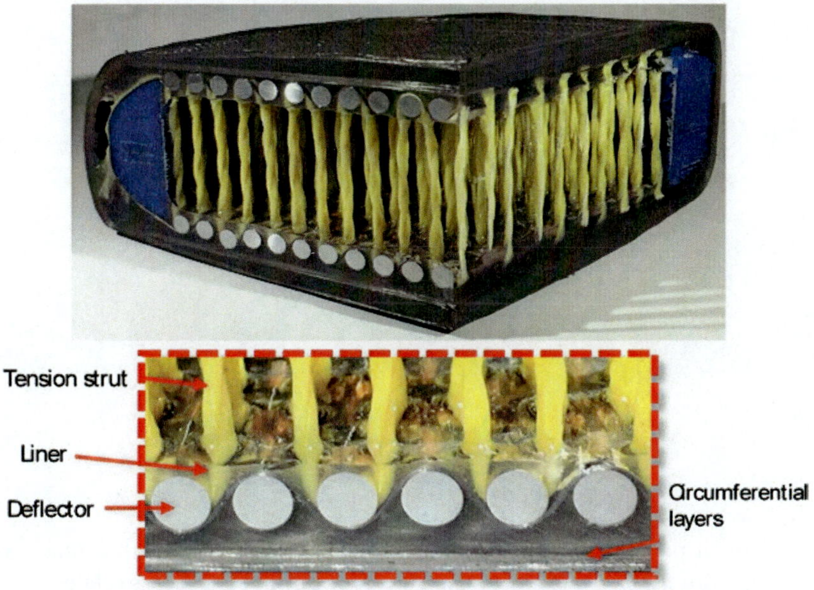

Fig. 10. Manufactured prototype of the 3D weaving in sectional view

4 Prototype testing

The prototypes were subjected to pressure loads to verify the load carrying capacity. They are pressurised by means of a hydraulic pump and oil as a medium until failure occurs. The first tests were performed with a reduced number of struts. This results in a strut failure in the middle of the tank although the design of the boss integration and the tank corners for the prototype were not finalized to sustain the full pressure load.

Figure 11 shows where the supporting tension struts are located in the test area., which has a cylindrical shape as a result to the roto moulding. The reduced number of struts provokes a strut failure and for the evaluation of the prototypes a Matlab routine was developed, which extrapolates the burst pressure to a fully equipped vessel [11]. This Matlab routine works on the basis of the following Eq. (1)

$$p_R = \frac{p_V * A_P}{n} * \frac{1}{A_S} * \frac{f_{realtank}}{f_{Prototype}} \tag{1}$$

p_R	Burst pressure: fully equipped tank
p_V	Burst pressure: prototype with reduced number of struts
A_P	Test area (prototype)
n	Number of supporting struts
A_S	Area per strut (real tank)
$f_{prototype}$	Filaments per strut (prototype)
$f_{realtank}$	Filaments per strut (real tank)

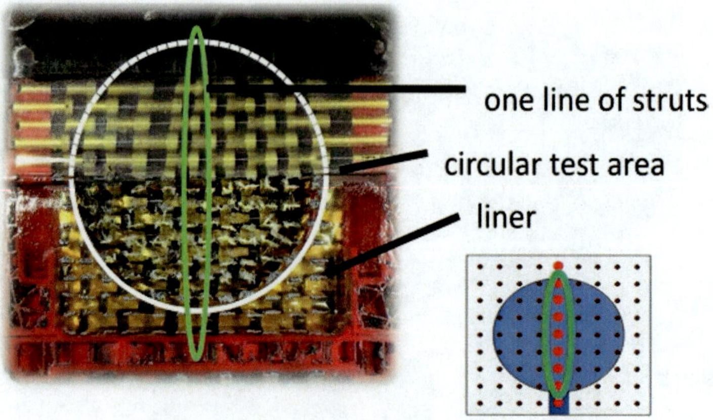

Fig. 11. Test area for pressure tests

Table 1 lists all parameters and results of the pressure tests of both prototypes. The higher number of filaments per strut ($f_{prototype}$) for tufting, is process related.

Table 1. Results of the extrapolation from the pressure tests

Parameter	Tufting	3D-weaving
p_V	160 bar	84 bar
A_P	1520 mm^2	1963 mm^2
n	11	14
A_S	16 mm^2	16 mm^2
$f_{Prototype}$	4000	2000
$f_{realtank}$	5000	5000
p_R: Burst pressure for a fully equipped tank	1860 bar	1980 bar

Figure 12 shows a sectional view of both prototypes and a microscopic view of the strut failure. In both prototypes the entire row of struts is cracked, which shows that all struts were exposed to the load from the pressure test. The results of the extrapolation show that both design principals are suitable for producing a tank that could withstand the required burst pressure with full strut occupation and appropriate strut material. These results must be validated in the further course of the project.

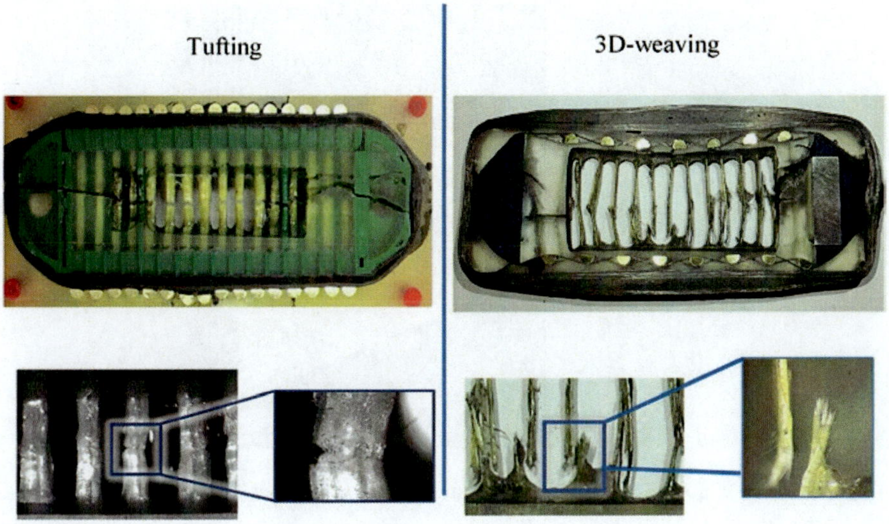

Fig. 12 Cross section of a pressure vessel after burst testing and microscope image of a broken tension strut

5 Industrialisation

5.1 Tufting process qualification for series production

To manufacture larger hydrogen pressure vessels there are critical Key technologies to be verified concerning the tufting process (e.g. needle-, looper- and core-design).

For an assumed tank height of 130 mm, a needle length of approximately 200 mm is required. The needle length is defined by a needle anti-buckling device (30 mm), a safety distance to the tank (10 mm), an installation height (130 mm) and a max. piercing for loop formation (30 mm). The company Groz-Beckert develop a prototype tufting needle with a length of 200 mm with a needle size of NM 250 (Fig. 13). Industrial tufting machines are state-of-the-art in rug production using needle bars with more than 100 needles and a production speed more than 1 m/min. However, due to this extremely high needle length, the transfer of the technologies for the tufting process developed for the strutted vessel is limited. New innovative tufting technologies must be developed for the industrialisation step as:

- Needle guidance
- Preparation of the aramid fibre
- Core technology
- Looper for deflector integration

Fig. 13. Needle bar and demonstrator in original Tank size

Figure 13 shows that manufacture using the tufting process is possible with an assumed tank height of 130 mm.

5.2 Weaving process qualification for series production

3D weaving is already state-of-the-art for composite materials in the case of spacer fabrics. With its TRITOS PP weaving machine the DORNIER company produces spacer fabrics in a serial process by means of Jacquard machines. However, these spacer fabrics are manufactured for sandwich constructions and are not suitable for internal pressure loads. The design of the standard machines will not fulfil the requirements, since the integration of the metallic deflector element during the weaving process is not covered. The developed prototype weaving machine at the Munich University of Applied Sciences can be used to study the process in detail and to verify design aspects for this new vessel type. The machine built in this way shows how a machine can be designed in its basic components to manufacture a pressure vessel using a complex 3D weaving process with loop deflectors. Key for a future series production is the transformation to the real tank dimensions. The basic principles of the machine remain the same even with larger vessel dimensions and scaling to a desired vessel size can be performed by a linear machine scaling. The vessel height depends on the size of the shafts and their travel distances, the vessel width is defined by the number of rovings for the weave structure and the vessel length is only limited by the length of the machine. In order to prove the suitability of the scaling method, the machine has been modified to build a demonstrator with the same height and length as the original tank (Fig. 14).

Fig. 14. Modified machine and demonstrator in original Tank size

6 Summary and Outlook

Two processes for integrating tension struts into a box-shaped vessel have been developed. The tufting process inserts the tension struts into the vessel from the outside, while the weaving process builds up the tension struts as an integral part of the tank wall. For both processes, special machines have been developed that can carry out these complex procedures and manufacture prototypes. The prototypes manufactured in this way serve to validate the design principles and have been tested in the first hydraulic pressure tests. Since the complexity of a pressure vessel at such high pressures cannot be represented with the current prototypes, prototypes with a lower number of struts have been manufactured and the results extrapolated to a fully equipped vessel. These results indicate that the target burst pressure of 157 MPa can be achieved with both manufacturing processes. Furthermore, the scalability to larger vessel dimensions has been explained and demonstrated. In the next step, the prototypes will be tested for their long-term tightness in cycle tests and the integration of the liner will be improved. The goal is to manufacture prototypes that can withstand the desired burst pressure and maintain their tightness. In addition, with the help of the two processes, sections of the vessel in real dimensions will be built in the form of demonstrators to validate the suitability of the machine technology for scalability. An additional aspect for future investigations is also the study of the load transfer between the woven structure and the circumferential carbon layers including the filler material.

Acknowledgements. The authors would like to express their gratitude to the German Federal Ministry for Economic Affairs and Energy for financial support within the government funded project "BRYSON".

References

1. European Commission, A European Green Deal: Striving to be the first climate-neutral continent. [Online]. Available: https://ec.europa.eu/info/strategy/priorities-2019-2024/europe-an-green-deal_en (accessed: Oct. 15 2021).
2. VDI/VDE, "Brennstoffzellen- und Batteriefahrzeuge: Bedeutung für die Elektromobilität," Düsseldorf, 2019. Accessed: Oct. 15 2021. [Online]. Available: https://www.vdi.de/ueber-uns/presse/publikationen/details/brennstoffzellen-und-batteriefahrzeuge
3. J. Töpler and J. Lehmann, Eds., *Wasserstoff und Brennstoffzelle: Technologien und Marktperspektiven,* 2nd ed. Berlin, Heidelberg: Springer Vieweg, 2017.
4. P. Komarnicki, J. Haubrock, and Z. A. Styczynski, *Elektromobilität und Sektorenkopplung: Infrastruktur- und Systemkomponenten,* 2nd ed. Berlin, Heidelberg: Springer Vieweg, 2020. [Online]. Available: http://www.springer.com/
5. M. Li et al., "Review on the research of hydrogen storage system fast refueling in fuel cell vehicle," *International Journal of Hydrogen Energy*, vol. 44, no. 21, pp. 10677–10693, 2019, doi: https://doi.org/10.1016/j.ijhydene.2019.02.208.
6. A. Kampker, D. Vallée, and A. Schnettler, Eds., *Elektromobilität: Grundlagen einer Zukunftstechnologie,* 2nd ed. Berlin: Springer Vieweg, 2018. [Online]. Available: http://swbplus.bsz-bw.de/bsz506370283cov.htm
7. BMW Group, Der erste BMW i4. [Online]. Available: https://www.press.bmwgroup.com/deutschland/article/detail/T0333329DE/der-erste-bmw-i4 (accessed: Oct. 15 2021).
8. United Nations Economic Commission for Europe (UN/ECE), "Regulation No. 134: Uniform provisions concerning the approval of motor vehicles and their components with regard to the safety-related performance of hydrogen-fuelled ve-hicles (HFCV)," 2015. Accessed: Oct. 15 2021. [Online]. Available: https://unece.org/fileadmin/DAM/trans/main/wp29/wp29regs/2015/R134e.pdf
9. *2021 Toyota Mirai Fuel Cell Vehicle | Innovation is Power.* [Online]. Available: https://www.toyota.com/mirai/ (accessed: Oct. 15 2021).
10. K. A. Öztas, R. A. Weerts, and M. G. Ruf, "An analytical and numerical approach to design a type I box-shaped pressure vessel with inner tension struts," *International Journal of Pressure Vessels and Piping*, vol. 192, p. 104441, 2021, doi: https://doi.org/10.1016/j.ijpvp.2021.104441.
11. M. Huber, A. Hupfeld, M. Ruf, and A. Horoschenkoff, "Bauraumoptimierter Wasserstofftank mit innerer Zugverstrebung," in *Leichtbau - von der Wissenschaft bis zur Schlüsseltechnologie für Ressourceneffizienz: 10. Landshuter Leichtbau-Colloquium : Tagungsband zum Colloquium : 24./25. Februar 2021, Hochschule Landshut*, O. Huber, M. Bicker, and P. Patzelt, Eds., Landshut: LC-Verlag, 2021.
12. M. Ruf *et al.,* "Neue Bauweisen Von Wasserstoffdruckbehältern Für Die Integration in Zukünftige Fahrzeugarchitekturen," in *Proceedings of the Munich Symposium on Lightweight Design 2020,* S. Pfingstl, A. Horoschenkoff, P. Höfer, and M. Zimmermann, Eds., Berlin, Heidelberg: Springer Berlin Heidelberg, 2021, pp. 74–85.

Multi-Objective Topology Optimization of Frame Structures Using the Weighted Sum Method

Martin Denk[1(✉)], Klemens Rother[2], Emir Gadzo[1], and Kristin Paetzold[3]

[1] Institute for Product Development, Universität Der Bundeswehr München, 85577 Neubiberg, Germany
denk.martin1990@gmail.com, emir.gadzo@unibw.de

[2] Institute of Material and Building Research, Munich University of Applied Science, 80335 Munich, Germany
klemens.rother@hm.edu

[3] Institute of Machine Elements and Machine Design, Technical University of Dresden, 01062 Dresden, Germany
kristin.paetzold@unibw.de

Abstract. Optimization of thermal compliance and linear elastostatic compliance is a common research task that can increase heat dissipation and structural stiffness. In particular, support structures in additive manufacturing (Kuo et al. in Struct. Multidiscip. Optim. 57:183–195, Jan. 2018) or the optimization of heat sinks (Zhou et al. in Struct. Multidiscip. Optim. 54:1045–1060, Oct. 2016) often require the consideration of both objectives. This article deals with the multi-objective topology optimization of frame structures using the weighted sum method for 2D and 3D wireframe meshes. Therefore, an element stiffness matrix is used, which couples both contributions. This automatically leads to holistic design proposals in terms of considering both objectives.

Keywords: Multi-objective · Topology Optimization · Unit Cell · Grid Structure

1 Introduction

Topology optimization attempts to achieve an optimal arrangement of material in a selected area or volume. As a rule, the material properties of individual elements of a finite element simulation are changed via a design variable (density). This design variable can scale material properties such as modulus of elasticity or thermal conductivity to simulate for example the removal of material in lightly strained domains and add material in highly strained domains for linear elastostatic use-case [3]. Multi-objective optimization is one of the recent challenges in topology optimization [3, 4],

J. Rieser et al. (Eds.): *Proceedings of the Munich Symposium on Lightweight Design 2021*, Proceedings, pp. 83–92, 2023. https://doi.org/10.1007/978-3-662-65216-9_8

where different objectives such as cooling a battery [5], improving the performance of solar energy [6], or reducing pressure losses [7] have to be considered. These objectives are often pursued independently or separately leading to suboptimal solutions [7, 8]. Therefore, the combination of such objectives in optimization is necessary, e.g., by using the weighted sum method [8, 9], which allows a simple approach to combining objective functions [9]. A coupled objective function can be defined by summing the individual objectives, scaled by a weight factor. In this work, the weighted sum is chosen due to the simplicity of the method, as it achieved reasonable results in previous publications with similar objective functions [11, 12]. For limitations and drawbacks of this weighted sum and alternative approaches, see [9].

Specific objectives, such as the simultaneous optimization of elastostatic and thermal compliance might be required for the design of masonry structures for example [10] or heat sinks that are also intended to withstand mechanical loads [11, 12]. Figure 1 shows the result of the approach used in this work to optimize a wireframe structure by minimizing thermal and elasostatic compliance while considering normalized weight factors w_s between zero and one. For the thermal and elastostatic case in the following example, a constant temperature and a fixed displacement (blue) are considered as boundary conditions. In addition, a constant volumetric heat source $\dot{Q}$ on the whole domain and a constant unit force (red arrow) are considered as load conditions. The weighting factor w_s considers the contribution of elastostatic compliance, while the contribution of thermal compliance is considered with $w_h = 1 - w_s$ in the opposite way. This concept is largely based on previous studies [11, 12], which supported a similar approach for 2D elements. The heat transfer case is based on the assumption of a constant temperature in the lower-left corner of the wireframe and volumetric heat source over the entire domain. The elastostatics case is defined by a fixed boundary condition at all nodes on the left of the wireframe and an applied force in the lower right corner.

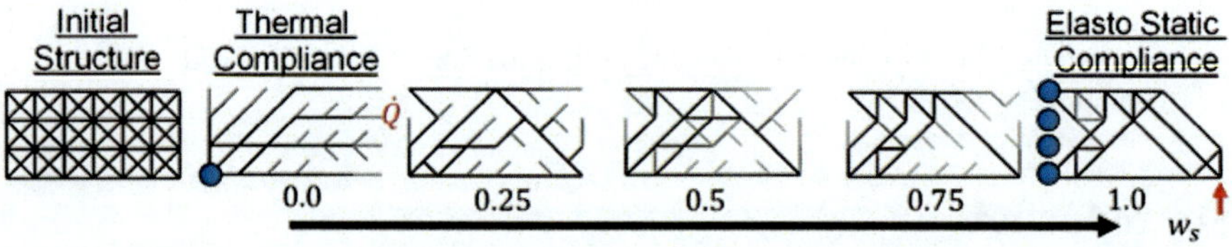

Fig. 1. Results for a multi-objective topology optimization using the weighted sum method for frame structures similar to [11, 12]. The heat transfer case is based on the assumption of a constant temperature in the lower-left corner of the wireframe and volumetric heat source over the entire domain. The elastostatic case is defined by a fixed boundary condition at all nodes on the left of the wireframe and a boundary condition in the lower right corner. The influence of thermal compliance decreases from left to right, while the influence of static compliance increases

This work focuses on applying multi-objective optimization to such frame- or lattice-type structures, where the reduction to a new element stiffness matrix greatly simplifies the material interpolation in topology optimization. The following state of the art part is briefly tailored to multi-objective optimization and wireframe structure optimization.

2 State of the Art

Optimization of frames and trusses or lattices is one of the most common tasks in structural optimization, where the geometric properties and the connections to the spar or beam elements are modified according to the objectives [3, 13]. Moreover, more advanced topics such as embedding nonlinearity [13], multi-objectivity [4, 14], flexible connections [15], or energy absorption [16] can be applied in topology optimization for frame structures. An objective such as thermal compliance or elastic compliance is widely used but often treated separately for frame structures [17].

The weighted sum method is applied by estimating the weighting factors using complete matrix configurations to optimize two objectives, similar to [11, 12]. In contrast to [11, 12], the whole definition of the multi-objective approach is embedded in the stiffness matrix so that material interpolation approaches such as solid isotropic material with penalization SIMP [18] can be used. For more details on the optimization of heat transfer systems, we refer in particular to the review articles [19] and [20]. Due to the sensitivities being always negative of the thermal and elastostatic compliance, the optimality criterion [3] can be applied in contrast to a genetic algorithm [4, 14]. An optimality criterion is a simplified approach in which the design variables of the optimization task are changed according to their sensitivity. Moreover, the topology of the lattice structure can be varied by using different lattice sizes [17].

3 Multi-Objective Element Stiffness Matrix in Topology Optimization

Multiple objective functions can be coupled using different approaches. In this work, a weighted sum is chosen. However, since the individual objective functions Π_j have different magnitudes, a normalization is necessary. Thereby, the individual objective functions are scaled by means of the objective values Π_j^* of so-called reference designs. This leads to comparable objectives and can be coupled by formulating and minimizing the total objective function Π_{Total} with

$$\min\left(\Pi_{\text{Total}} = \sum\nolimits_{j=1}^{u} w_j \frac{\Pi_j}{\Pi_j^*}\right), \tag{1}$$

where u describes the number of different objectives and w_j the corresponding weight factor of the objective Π_j. A suitable way to scale the individual objective contributions is the calculation of the objective of a previous simulation with full material properties Π_j^0 (volume ratio of 1.0). Each element is defined with unscaled modulus of elasticity or thermal conductivity values, which are used to calculate the mean compliance. By coupling two objectives using the element stiffness matrix for thermal conduction $\boldsymbol{k}_h$ and elastostatics $\boldsymbol{k}_s$ for k elements with

$$\Pi_{\text{Total}} = w_s\Pi_s + w_h\Pi_h = \frac{w_s}{\Pi_s^0}\sum\nolimits_{j=1}^{k} \boldsymbol{u}_j^T \boldsymbol{k}_{s,j} \boldsymbol{u}_j + \frac{w_h}{\Pi_h^0}\sum\nolimits_{j=1}^{k} \boldsymbol{T}_j^T \boldsymbol{k}_{h,j} \boldsymbol{T}_j \tag{2}$$

the influence of the thermal compliance and elastostatic compliance using the displacements $\boldsymbol{u}_j$ and temperatures $\boldsymbol{T}_j$ of one element j can be weighted by w_s and w_h independently. The weighted contributions can now be assembled to the multi-objective element stiffness matrix for a 2D-beam $\boldsymbol{k}_x$ by

$$\begin{bmatrix}
\frac{w_h}{\Pi_h^0}\frac{A\lambda}{2a} & 0 & 0 & 0 & -\frac{w_h}{\Pi_h^0}\frac{A\lambda}{2a} & 0 & 0 & 0 \\
 & \frac{w_s}{\Pi_s^0}\frac{AE}{2a} & 0 & 0 & 0 & -\frac{w_s}{\Pi_s^0}\frac{AE}{2a} & 0 & 0 \\
 & & \frac{w_s}{\Pi_s^0}\frac{3EI_z}{2a^3} & \frac{w_s}{\Pi_s^0}\frac{3EI_z}{2a^2} & 0 & 0 & -\frac{w_s}{\Pi_s^0}\frac{3EI_z}{2a^3} & \frac{w_s}{\Pi_s^0}\frac{3EI_z}{2a^2} \\
 & & & \frac{w_s}{\Pi_s^0}\frac{2EI_z}{a} & 0 & 0 & -\frac{w_s}{\Pi_s^0}\frac{3EI_z}{2a^2} & \frac{w_s}{\Pi_s^0}\frac{EI_z}{a} \\
 & & & & \frac{w_h}{\Pi_h^0}\frac{A\lambda}{2a} & 0 & 0 & 0 \\
 & & & & & \frac{w_s}{\Pi_s^0}\frac{AE}{2a} & 0 & 0 \\
 & sym. & & & & & \frac{w_s}{\Pi_s^0}\frac{3EI_z}{2a^3} & -\frac{w_s}{\Pi_s^0}\frac{3EI_z}{2a^2} \\
 & & & & & & & \frac{w_s}{\Pi_s^0}\frac{2EI_z}{a}
\end{bmatrix}$$

where A is the cross-sectional area, λ is the conductivity, E is the modulus of elasticity, a is the half-length of the element, and I_z is the second moment of area corresponding to z. The vector of degrees of freedom can be composed by embedding the temperature T and the deformations $[u, v, \phi_z]$ of the first and second nodes of a 2D-beam to

$$\boldsymbol{x} = \left[T_1, u_1, v_1, \phi_{z,1}, T_2, u_2, v_2, \phi_{z,2},\right], \tag{3}$$

which gives the expression for the weighted sum with

$$\min\left(\Pi_{Total} = \sum\nolimits_{j=1}^{k} \boldsymbol{x}_j^T \boldsymbol{k}_{x,j} \boldsymbol{x}_j\right). \tag{4}$$

The matrix $\boldsymbol{k}_x$ can now be applied to the topology optimization using isotropic material with penalization (SIMP) [21] with the penalty exponent p as

$$\boldsymbol{k}_{x,j} = \rho_j^p \boldsymbol{k}_{x,0}. \tag{5}$$

where $\boldsymbol{k}_{x,0}$ describes the element stiffness matrix with full conductivity and elasticity modulus, ρ_j the normalized density of element j and $\boldsymbol{k}_{x,j}$ the scaled element stiffness matrix referred to element j. This approach is applied to the topology optimization of single objectives, with the corresponding weighting factors $w_{min} = 0.01$ close to zero in the following section. To demonstrate the universal concept of the multi-objective element stiffness matrix $\boldsymbol{k}_x$, topology optimization is also performed using the so-called bidirectional evolutionary structural optimization (BESO) and different initial structures for the wireframe mesh.

4 Single Objective Topology Optimization

Especially for topology optimization, a 0–1 density field is desirable when using a normalized density [3, 22]. Otherwise, an interpretation of the greyscale results (uncertain material distribution) is required. To measure this uncertainty, the binary cross-entropy

$$m_c = \frac{1}{n}\sum\nolimits_{i=1}^{n} - \left(\rho_i \log_a(\rho_i) + (1-\rho_i)\log_a(1-\rho_i)\right) \tag{6}$$

can be used as a criterion in topology optimization [12], where n is the number of design variables and a is the base of the logarithm, often chosen as two. Such a measure is zero when a unique 0-1density field emerges and to a high value when the density is continuously distributed. For the mesh to start with, a rectangular grid is chosen. Based on that initial design structure, 1D elements can be removed or added during the optimization. By selecting a volume ratio with 30%, only that ratio can be filled with material, measured by the summation of the density times the element length. Figure 2 shows the optimization of a load case minimizing the elastostatic compliance. First, the initial structure and the boundary and load conditions are visualized under a). Then the topology optimization is performed, whose relative objective function value and entropy decrease initially and converge after several iterations in b). The volume ratio is almost constant during the optimization. The final solution now consists of elements with variable material distribution in c) hiding elements with density close to zero.

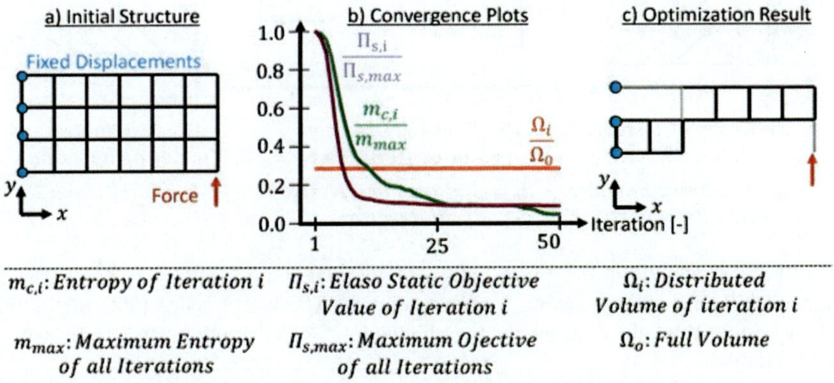

Fig. 2. Topology optimization minimizing elastostatic compliance using SIMP with $w_h = 0$: a) Initial structure and applied load case. b) Convergence plots for the relative objective $\Pi_{s,i}/\Pi_{s,max}$ relative entropy $m_{c,i}/m_{c,max}$ and volumetric ratio Ω_i/Ω_0. c) Result using a volume ratio of 0.3

The case in Fig. 2 shows, that this optimization leads to a density of zero or one for most elements, so that the uncertainty in terms of entropy is small. Due to the limited topology of the initial structure, the elements can only be applied in x- or y-direction, making it difficult to support the load in the given load case.

As an alternative, evolutionary structure optimization (ESO) [23] or its improved algorithm, bidirectional evolutionary structure optimization (BESO), can be used, which continuously removes inefficient material by reducing the allowable volume ratio. Such optimization can also be used for various optimization tasks, such as considering multi-objective or multi-load-case [22]. The initial structure can also be modified to add 45° beams. Figure 3 shows the results of applying the BESO implementation described in [22] to minimize the thermal compliance with the new initial structure a) the corresponding convergence plot b) and the optimization result c). In contrast to the solution in Fig. 2 with SIMP, the value of the thermal compliance increases (yellow line). In the first iteration, the lowest objective function value is reached due to the material ratio of one in the entire model, which steadily increases due to the removal of the material. The material is linearly removed until the 37th iteration and then kept constant. In the last segment with a constant volume ratio, the objective value first decreases and then converges. The abrupt change in the assignment of the material can create separated domains. Since the density values in BESO can only assume zero and one, regions with slightly different sensitivities are defined, one with full thermal conductivity and the other with minimum thermal conductivity. So these slightly different sensitivities can lead to the gap shown in Fig. 3.

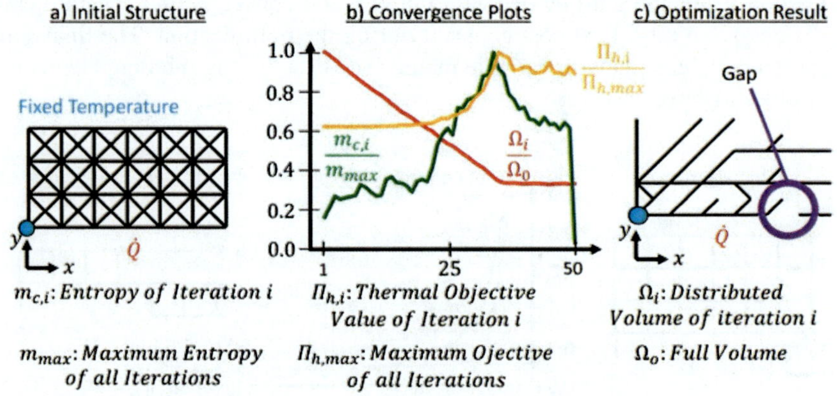

Fig. 3. BESO strategy for the optimization of thermal compliance $w_s = 0$. a) The initial structure with a unit cell consisting a center point. b) The convergence plots of the thermal compliance (objective), the entropy and the volume ratio. c) The optimization result with a gap in the structure

The minimization of thermal compliance can lead to heat exchanger designs that transfer heat from the fixed temperature to the environment as uniformly as possible (minimization of the mean temperature). This can lead to a fan-like grid structure that distributes the temperature as evenly as possible in the domain.

In general, BESO leads to a binary decision after each iteration, but to achieve better convergence, an averaging of the density of the current iteration and the previous iteration is performed in [22]. This density is then used to scale the material properties such as thermal conductivity or elasticity. The density is then converted back to a binary density field. The entropy plot in Fig. 3 shows the entropy of the averaged densities. Since there is no more averaging in the last step, the entropy drops to 0. BESO or ESO, unlike SIMP, lead to a 0–1 design (densities), which is often desirable. SIMP leads to a uniform decrease in the objective function over the iterations compared to BESO. Therefore, both strategies are often used in topology optimization [22]. Similar results can now be obtained by using nonzero weighting factors for both objectives.

5 Multi-Objective Topology Optimization of Frame Structures

The same element stiffness matrix can now be applied to the above-mentioned thermal case and elastostatic case by changing the weighting factors. Figure 1 shows the variation of the weighting factors using a unit cell with center node and SIMP. Visually, the influence of elastostatic compliance steadily decreases with decreasing weighting factor w_s, while the impact of thermal compliance $w_h = 1 - w_s$ increases.

The same concept can now be applied to a 3D load case. Figure 4 shows a) the selected 3D mesh (cubic-space-centered unit cell) and the boundary (blue, green), and load conditions (red). In particular, the constant volumetric heat source is applied to each element. Figure 4 b) shows the convergence graphs for the relative entropy, volume ratio and two objective functions for each iteration *i*.

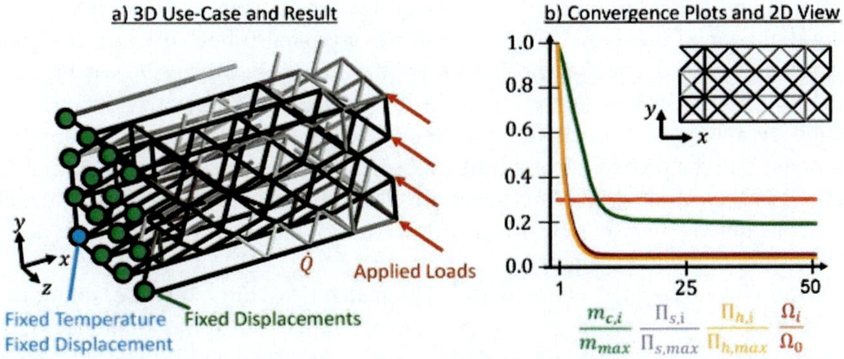

Fig. 4. Multi-objective optimization of a 3D frame structure using weighted sum using equal weight factors $w_s = w_h = 0.5$: a) 3D mesh and load case and optimized result using SIMP. b) Convergence plots of the thermal compliance, elasto static compliance, entropy and volume ratio

The visual result shows a holistic design, which is also reflected in the convergence of the graphs. Both objective function values decrease steadily and converge at the end. The entropy in the example is very high compared to Fig. 2, which is reflected by many grey elements, especially in the 2D view.

The presented multi-objective element stiffness matrix can solve coupled heat conduction problems and elastostatic problems. This concept can be extended in convection or fluid mechanics in future work. Similarly, the presented method can now be used for real-world applications, such as passive cooling of a CPU and considering contact pressure. Another application would be the simultaneous passive cooling of a battery in a vehicle, taking into account the mechanical loads. The wireframe should be designed to simultaneously cool and sustain mechanical loads.

The main drawback of the presented multi-objective optimization is the simplicity of the physical model. Only linear elasticity and the heat conduction equation are chosen. If nonlinear material behavior and more complex tasks such as turbulent flow are considered, finding normalization values can be challenging. Similarly, it is then questionable whether the weighted sum leads to a Pareto front [9]. In addition, convection using a constant volumetric heat source is greatly simplified.

6 Summary

This article presents a multi-objective topology optimization for lattice-type structures using the weighted sum method. An element stiffness matrix is proposed in which heat transfer and elastostatic functionality is embedded and the corresponding weighting factors are included. This element stiffness matrix was first applied to single-factor topology optimization for thermal and elastostatic compliance for 2D wireframes and applied for a multi-objective case of a 3D wireframe.

Minimizing the mean thermal compliance leads to a heat sink design while minimizing the mean elastostatic compliance shows a typical frame structure design such as cranes or bridges. The weight factors for the multi-objective 3D (see Fig. 2) and 2D (see Fig. 1) optimization were chosen to consider both objective functions. The resulting geometry withstands the mechanical loads (mean elastostatic compliance) while satisfying the properties of a heat sink (mean thermal compliance). A significant drawback is that these results are uncertain if solid isotropic material with penalization is considered. By using BESO such uncertainty can be avoided, which does not lead to a smoothly decreasing objective function.

By defining a coupled element stiffness matrix, existing software solutions can be used directly by only considering that a new element has to be implemented. This approach reduces the implementation effort in contrast to a multi-objective optimization shown in [11].

These optimized wireframes can now be further investigated using an experimental setup. Moreover, this approach can be extended in further articles by embedding objectives such as fluid flow similar to [12] or by using non-uniform unit cells similar to [17].

References

1. Y.-H. Kuo, C.-C. Cheng, Y.-S. Lin, and C.-H. San, 'Support structure design in additive manufacturing based on topology optimization', Struct. Multidiscip. Optim., vol. 57, no. 1, pp. 183–195, Jan. 2018, doi: https://doi.org/10.1007/s00158-017-1743-z.
2. M. Zhou, J. Alexandersen, O. Sigmund, and C. B. W. Pedersen, 'Industrial application of topology optimization for combined conductive and convective heat transfer problems',

Struct. Multidiscip. Optim., vol. 54, no. 4, pp. 1045–1060, Oct. 2016, doi: https://doi.org/10.1007/s00158-016-1433-2.

3. M. P. Bendsoe and O. Sigmund, Topology Optimization: Theory, Methods, and Applications, 2nd ed. Berlin Heidelberg: Springer-Verlag, 2004. doi: https://doi.org/10.1007/978-3-662-05086-6.
4. J. Lim, C. You, and I. Dayyani, 'Multi-objective topology optimization and structural analysis of periodic spaceframe structures', Mater. Des., vol. 190, p. 108552, May 2020, doi: https://doi.org/10.1016/j.matdes.2020.108552.
5. A. Chu, Y. Yuan, J. Zhu, X. Lu, and C. Zhou, 'The Design and Investigation of a Cooling System for a High Power Ni-MH Battery Pack in Hybrid Electric Vehicles', Appl. Sci., vol. 10, p. 1660, 2020, doi: https://doi.org/10.3390/app10051660.
6. D. K. Gupta, M. Langelaar, M. Barink, and F. van Keulen, 'Topology optimization of front metallization patterns for solar cells', Struct. Multidiscip. Optim., vol. 51, no. 4, pp. 941–955, Apr. 2015, doi: https://doi.org/10.1007/s00158-014-1185-9.
7. E. Dede, 'Multiphysics Topology Optimization of Heat Transfer and Fluid Flow Systems', 2009.
8. K. Proos, G. Steven, O. Querin, and Y. Xie, 'Multicriterion evolutionary structural optimization using the weighting and the global criterion methods', AIAA J., vol. 39, pp. 2006–2012, 2001, doi: https://doi.org/10.2514/3.14961.
9. R. T. Marler and J. S. Arora, 'The weighted sum method for multi-objective optimization: new insights', Struct. Multidiscip. Optim., vol. 41, no. 6, pp. 853–862, Jun. 2010, doi: https://doi.org/10.1007/s00158-009-0460-7.
10. G. Vantyghem, W. Corte, V. Boel, and M. Steeman, 'Structural and thermal performances of topological optimized masonry blocks', presented at the Asian Congress of Structural and Multidisciplinary Optimization 2016, Nagasaki, Japan, 26-05 2016.
11. M. Denk, K. Rother, and K. Paetzold, 'Multi-Objective Topology Optimization of Heat Conduction and Linear Elastostatic using Weighted Global Criteria Method', in Proceedings of the 31st Symposium Design for X (DFX2020), Bamberg, Sep. 2020, vol. 31, pp. 91–100. doi: https://doi.org/10.35199/dfx2020.10.
12. M. Denk, K. Rother, M. Zinßer, C. Petroll, and K. Paetzold, 'Nodal cosine sine material interpolation in multi objective topology optimization with the global criteria method for linear elasto static, heat transfer, potential flow and binary cross entropy sharpening.', in Proceedings of the 23th International Conference on Engineering Design, Götheburg, Aug. 2021, vol. 1, pp. 2247–2256. doi: https://doi.org/10.1017/pds.2021.486.
13. N. Changizi and G. P. Warn, 'Topology optimization of structural systems based on a nonlinear beam finite element model', Struct. Multidiscip. Optim., Jul. 2020, doi: https://doi.org/10.1007/s00158-020-02636-x.
14. J. N. Richardson, S. Adriaenssens, P. Bouillard, and R. Filomeno Coelho, 'Multiobjective topology optimization of truss structures with kinematic stability repair', Struct. Multidiscip. Optim., vol. 46, no. 4, pp. 513–532, Oct. 2012, doi: https://doi.org/10.1007/s00158-012-0777-5.
15. H. Fredricson, T. Johansen, A. Klarbring, and J. Petersson, 'Topology optimization of frame structures with flexible joints', Struct. Multidiscip. Optim., vol. 25, no. 3, pp. 199–214, Aug. 2003, doi: https://doi.org/10.1007/s00158-003-0281-z.
16. C. B. W. Pedersen, 'Topology optimization design of crushed 2D-frames for desired energy absorption history', Struct. Multidiscip. Optim., vol. 25, no. 5–6, pp. 368–382, Dec. 2003, doi: https://doi.org/10.1007/s00158-003-0282-y.
17. Y. Han and W. F. Lu, 'A Novel Design Method for Nonuniform Lattice Structures Based on Topology Optimization', J. Mech. Des., vol. 140, no. 9, Jul. 2018, doi: https://doi.org/10.1115/1.4040546.

18. M. P. Bendsøe and O. Sigmund, 'Material interpolation schemes in topology optimization', Arch. Appl. Mech., vol. 69, no. 9, pp. 635–654, Nov. 1999, doi: https://doi.org/10.1007/s004190050248.
19. J. Alexandersen and C. S. Andreasen, 'A Review of Topology Optimisation for Fluid-Based Problems', Fluids, vol. 5, no. 1, Art. no. 1, Mar. 2020, doi: https://doi.org/10.3390/fluids5010029.
20. T. Dbouk, 'A review about the engineering design of optimal heat transfer systems using topology optimization', Appl. Therm. Eng., vol. 112, pp. 841–854, Feb. 2017, doi: https://doi.org/10.1016/j.applthermaleng.2016.10.134.
21. Y. M. Xie and G. P. Steven, 'A simple evolutionary procedure for structural optimization', Comput. Struct., vol. 49, no. 5, pp. 885–896, Dec. 1993, doi: https://doi.org/10.1016/0045-7949(93)90035-C.
22. X. Huang and Y. Xie, 'Evolutionary Topology Optimization of Continuum Structures: Methods and Applications', John Wiley Sons, pp. 121–150, 2010, doi: https://doi.org/10.1002/9780470689486.ch7.
23. Y. Xie and G. Steven, 'Shape and layout optimization via an evolutionary procedure', presented at the International Conference on Computational Engineering Science, Hong Kong, Dec. 1992.

Parametrization of Cross-Sections by CNN Classification and Moments of Area Regression for Frame Structures

Martin Denk[1(✉)], Klemens Rother[2], Josef Neuhäusler[2], Christoph Petroll[3], and Kristin Paetzold[4]

[1] Institute for Product Development, Universität Der Bundeswehr München, 85577 Neubiberg, Germany
denk.martin1990@gmail.com

[2] Institute of Material and Building Research, Munich University of Applied Science, 80335 Munich, Germany
{klemens.rother,josef.neuhaeusler}@hm.edu

[3] The Bundeswehr Research Institute for Materials, Fuels and Lubricants, Universität Der Bundeswehr München, 85435 Erding, Germany
christophpetroll@bundeswehr.org

[4] Institute of Machine Elements and Machine Design, Technical University of Dresden, 01062 Dresden, Germany
kristin.paetzold@unibw.de

Abstract. The shape reconstruction of volumetric images can require classifying cross-sections such as circles or rectangles. Depending on the type of cross-section, shape parameters such as radius, width, or height must be regressed. This article addresses cross-section classification by convolutional neural networks (CNN), with further regression of the shape parameter using the moments of area. This fully recognizes the cross-section type and the corresponding geometric shape parameter for 2D binary images.

Keywords: Reverse Engineering · Artificial Intelligence · Cross-Sections · Neural Networks

1 Introduction

Applications for a medical purpose [1, 2] or structural optimization [3, 4] often require the parametrization of surfaces. Such parametrization can be covered using primitive surface types [5, 6], freeform surfaces [7, 8] or its combination. In reverse engineering of 3D images resulting from computer tomography (CT) scans [1] or topology optimization [9, 10], polygon meshes [7] or point clouds [11] might be given.

Such geometric representations can be segmented into patches [5, 6], or parts [12, 13], which are used to find explicit or implicit shape descriptions. One common

J. Rieser et al. (Eds.): *Proceedings of the Munich Symposium on Lightweight Design 2021*, Proceedings, pp. 93–103, 2023. https://doi.org/10.1007/978-3-662-65216-9_9

segmentation algorithm followed by surface parametrization can be covered using curve [14] or surface skeletons [15]. Particularly for frame-like structures, curve skeletons can lead to beamlines [14, 16]. These curve skeletons serve as a shape descriptor for frame-like organic geometries [17], which can be applied for deformed Finite-Element(FE)-meshes represented in [7], point cloud reconstruction [11], the determination of center lines in CT-Data to reconstruct blood vessels [1], vascular skeletons [2], topology optimization results [3, 4] or fast concept design [16]. While such structures are dedicated to a circular and elliptical cross-sections, non-circular shapes can also be processed. Figure 1 shows a) a car's side panel represented as a 3D image consisting of voxels. This must be converted into its parametric representation to be further processed, e.g. by specifying position and shape tolerances. By slicing the geometry along the beamlines and then estimating the shape parameters, sections shown in b) are generated. These sections represent 2D images for which c) parametric representations need to be estimated.

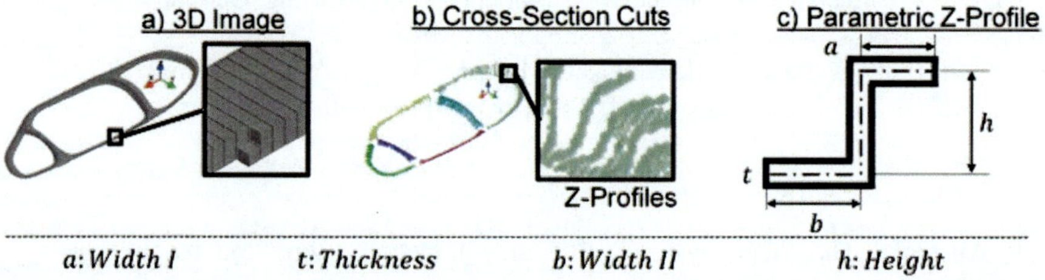

Fig. 1. Motivation for reconstructing a side panel of a car. a) 3D image resulting from a computed tomography scan or rasterization of a polygon mesh. b) Cross-section along the frame segments. c) The parametric representation of a Z-profile in its shape parameters

For automation, the type of cross-section must first be identified. Then, the shape parameters can be estimated. This article presents a hierarchical approach to automatically classify and regress such shape parameters from 2D images. The classification is solved by using convolutional neural networks and the regression is solved by multi-objective optimization by considering moments of area. The following chapter covers the state of the art of reconstruction in topology optimization and image classification for binary cross-sections.

2 State of the Art

In most reconstruction procedures that use skeletonization, the skeleton serves as a guide for the cross-sections or segmentations. Once a beamline is recognized, the corresponding cross-section can be determined. One standard approach is to define spherical or elliptical cross-sections manually, particularly for topology optimization results [18]. More convenient is the automatic determination of the cross-section parameters [3, 4, 19, 20], in which, for example, the radius is estimated via the distance of the beamline to the surface [3, 20]. Similar to cross-sections defined by geometrical primitives, B-spline cross-sections can be considered manually to cover the shape of the result from topology optimization [21]. Alternatively, it is possible to determine the parameters of such a freeform curve by geometric approximation [14, 16] or by considering a fit of the contour to a cross-section [22]. It is common to assume a circular cross-section due to the organic shape of topology optimization results. To be able to consider arbitrary cross-sections, this work first classifies the individual cross-section types in such a way that, in addition to circular cross-sections, common shapes such as the Z-section in Fig. 1 can also be considered.

The method from the previous work in [22] is used in the following for the classification of the cross-sections. By using a neural network the cross-section types of 2D images are automatically determined. To reduce the data variation, some feature engineering steps were used. These take advantage of the property that the cross-section type does not change due to scaling, shifting and rotation. Some of these invariants can be incorporated into the data set in a way that, for example, two different-sized squares can be reduced to one data sample. Particularly, to achieve rotational invariance, all cross-sections are rotated in the principal axes, resulting in four possible rotations (on the coordinate axes) that have the same principal axes. The individual transformed images are then color-coded based on the Euclidean distance of each pixel from the boundary edges. Figure 2 shows an example of such a Z-profile a) rotated in the principal axes, to which b) the distance field is applied. This distance field can also be calculated for the background by inverting the binary image. Finally, the c) distance field of the background is subtracted from the foreground, d) resulting in a final color coding of the entire image space. Based on this color coding, the individual cross-sections can now be classified.

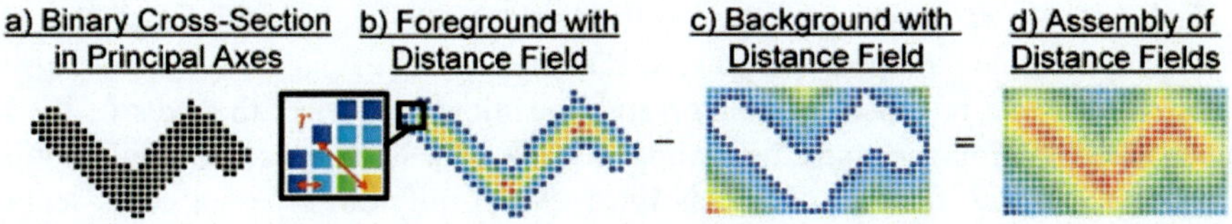

Fig. 2. Feature engineering of a Z cross-section applied according to [23]: a) The binary image is rotated into its principal axes. b) The distance of each pixel to its boundary is calculated for the foreground. c) The distance is also calculated for the background (inverted image). d) Result generated by subtracting the distance field of the foreground from the background

This work performs a regression based on the classification from [22], resulting in a hierarchical architecture. A given image is transformed according to the abovementioned methods and then classified. After the class is known, the corresponding shape parameters are estimated. The following section describes the methods chosen for this hierarchical classification followed by regression.

3 Hierarchical Classification and Regression of Cross-Sections

The synthetic dataset is created the same way as described in [23], but in contrast, the required data is halved by adding another transformation step. The data set is generated by creating individual polygons from randomly selected shape parameters. Initially, a polygon is created, e.g., the outer edges of a Z-profile, which is then converted into a binary pixel image by rasterization. This pixel image is now rotated into the principal axes using the transformations described above, scaled into a bounding box, and moved to the lower-left corner of the bounding box. These image transformations are not lossless because the rotated, shifted or scaled image no longer fits the original grid. In this work, bilinear resampling is used for this purpose, where the original binary image is interpolated into the newly transformed raster. The interpolation results lead to grey scaled areas, which can be converted into a binary image considering a greyscale limit (in this work 50%). All grey values above that limit are considered black, while all values below are considered white. Finally, the distance transformation is applied to the binary image, resulting in color coding as described above. In contrast to [23], the coordinates of the image are swapped based on the respective moments of area. First, it is checked whether the greater principal axis moment of the image $I_1^P > I_2^P$ corresponds to the moment of area I_y^P of the image. If this is not the case, the coordinate axes of the images are swapped. In this way, it can be ensured that e.g. the longer side of a rectangle is aligned along the y-axis. Since the mirroring of a cross-section about a coordinate axis does not change the moments of area, four possible orientations with the same moments of area result. Altogether, for the data argumentation the data set is only quadrupled, in which a data sample is mirrored about the coordinate axis y, z, y followed by z.

Each of these transformation steps is then stored in a database. Since the regressed parameters are calculated for the transformation invariant images, they must also be rotated, scaled and returned to their original position. Likewise, the axes of the previously changed images must be swapped back. These images are applied to the convolutional neural network that classifies several different types of cross-sections. Figure 3 illustrates a) the architecture for classification presented in [23], b) the moments of area computation of a binary image presented in [23], and furthermore c) the hierarchical architecture for the regression task presented in this work.

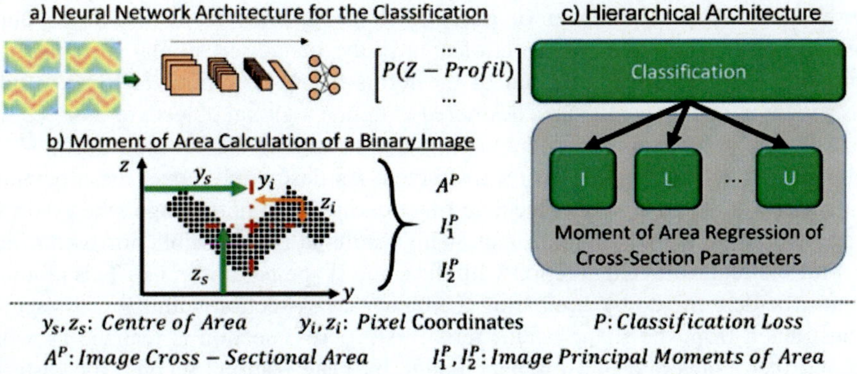

Fig. 3. Hierarchical classification and shape parameter regression of the cross-sections. a) Architecture of classification using four rotated representations of the distance fields. b) Moment of area calculation of a binary image for parametric cross-section regression. c) Hierarchical architecture using cross-section type classification followed by tailored regression

Based on the result of the classification, a regression dependent on the cross-section type can be performed. In this work, the regression of the shape parameter can be covered by comparing the principal moments of area of the image I_1^P, I_2^P with the principal moments of area of the parametric representation I_1, I_2. Furthermore, the cross-sectional areas of the image A^P and the parametric description A must match, which leads overall to a multi-objective optimization task by

$$\min\left(\left|I_1 - I_1^P\right|; \left|I_2 - I_2^P\right|; \left|A - A^P\right|\right). \tag{1}$$

The shape parameters x_i are constrained between non-negative and the maximum possible diagonal length of the image in the bounding box l_{max} to $x_i \in [0, l_{max}]$ by

$$l_{max} = \begin{cases} \sqrt{2}b_y \; for \; b_y \geq b_z \\ \sqrt{2}b_z \; for \; b_z > b_y \end{cases}, \tag{2}$$

where b_y describes the length of the bounding box along the y-axis and b_z describes the length of the bounding box along the z-axis. Instead of comparing the moments of area directly with each other, a reasonable normalization of the individual terms can be used to achieve a uniform ratio [24]. Likewise, instead of the sepearted objectives in (1), a smooth differentiable function over the error squares should be used to be able to find gradients for the optimization. From this, the formulation of the multi-objective optimization problem for the shape parameter x_i with the respective constraint can be derived as

$$\min\left(\left(\frac{A(x_0, \ldots, x_n)}{A^P} - 1\right)^2 + \sum\nolimits_{l=1}^{2}\left(\frac{I_l(x_0, \ldots, x_n)}{I_l^P} - 1\right)^2\right) \tag{3}$$

$l_{max} > x_i > 0,$

where n describes the number of possible shape parameters in the corresponding cross-section. Thus, if the moments of area in the image are similar to those of the parametric representation, the sum of the terms in (3) is minimized. The weighted sum with equal weights in (3) is considered to define a global objective.

If the shape parameters x_i are approximated according the solution of Eq. (3), the individual cross-sections can be reconstructed. As described above, the moments of area and thus the value of the objective function in (3) do not change when the shape parameters are mirrored along the axes. This results in four possible mirrored orientations for the reconstructed solution with the same shape parameters x_i. This alignment can now be determined by comparing the binary cross-section with the cross-sections reconstructed from the shape parameters. First, a cross-section is realized as a polygon and then converted into a binary image by rasterization. Second, the rasterized image is moved to the center of area of the original image. Third, this image is mirrored at the coordinate axes a) y, b) z c) y followed by z. These four images I_s in matrix representation of size mxn are now compared with the original image I_P in the matrix representation of size mxn by calculating the difference ϵ by summing up the individual pixels (black = 1, white = 0) using

$$\epsilon = \sum\nolimits_{y=0}^{m} \sum\nolimits_{z=0}^{n} |I_s(y,z) - I_P(y,z)|. \tag{4}$$

This results in four individual metrics ϵ for the four orientations leading to the final orientation with the smallest difference. Finally, the shape parameters are shifted, scaled and rotated back to the original position. If the axes have been swapped, they will be changed back. In the following section, the chosen regression approach is discussed through a case study.

4 Case Study on Classification and Regression

Initially, the synthetic dataset is trained for classification using the architecture of [23]. Since a different number M of cross-section types is used, the categorical cross entropy is for the sample i of n data samples considered as a loss function (as metric of accuracy) applying

$$L = -\sum\nolimits_{i=1}^{n} \sum\nolimits_{j=1}^{M} \widehat{y}_{i,j} log\left(y_{i,j}\right), \tag{5}$$

where $y_{i,j}$ describes the predicted probability for class j and $\widehat{y}_{i,j}$ the target probability value for class j [23]. For the encoding of the class vector a one-hot encoding is used, so that in a vector $\widehat{\mathbf{y}}_i$ of length M the value 1 is assigned to the associated class and the value zero to the remaining entries. The neural network now tries to predict $\widehat{\mathbf{y}}_i$ by estimating the class probabilities for $\mathbf{y}_i$ to minimize (5). Since four mirrored alternatives are possible for a single image, the individual probabilities are added and averaged. This results in an average probability from which the final class is derived. Overall, the ratio between correctly classified samples and all samples is 94%. Figure 4 shows a) the progression of accuracy over each training epoch, b) an example of averaging the four variables, and c) the composition of the vector for each category i.

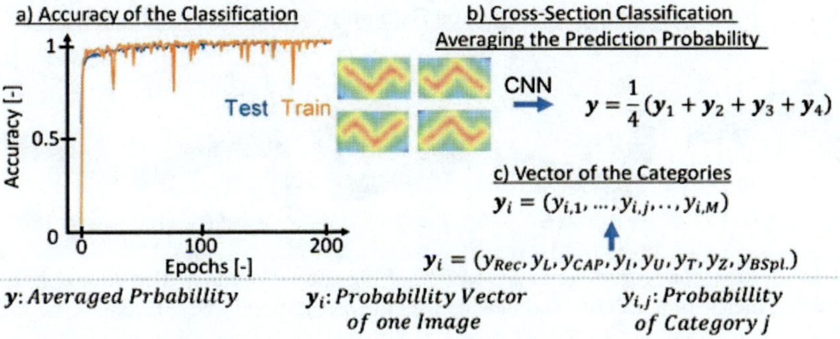

Fig. 4. Classification of cross-section type using categorical cross-entropy. a) Loss values during training. b) Averaging of the four probabilities into a general probability vector. c) Sampling of the probability vector considering the different cross-section classes

The full data set is separated into 80% training data and 20% test data. The accuracy of the training data set and the test data set converges towards 94%, while the accuracy of the test data set is usually lower than that of the training data set. This classification can now be tested on different examples. Figure 5 shows the classification result of various binary images, with the highest probabilities above 25% listed below the example.

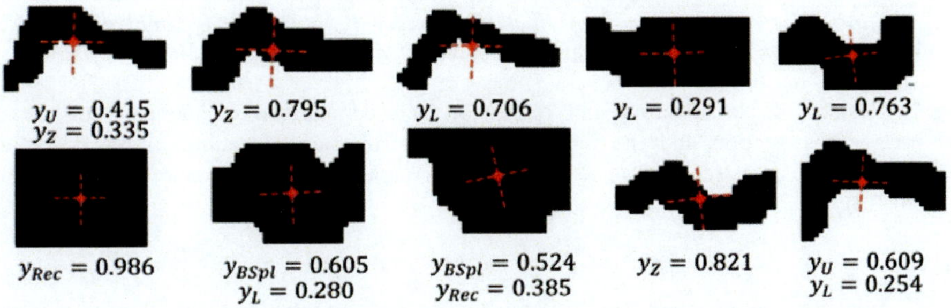

Fig. 5. Results of classification with a probability in y_i above 0.25, where $y_{[]}$ represents the probability of the corresponding cross-section type such as rectangle "Rec", B-spline "BSpl" or U-, L- Z-sections "U", "L", "Z" for sample i

The classification of the proposed samples is even for humans challenging due to the unclear shape. The first example shows a shape that could be classified as a "U", "Z" or "L" section. The neural network predicts a "U" or "Z" profile as possible configurations. Nevertheless, the classification usually results in a number of possible alternatives offered to the user.

Based on this classification, a regression of the shape parameters can now be performed by solving the optimization task (33), which requires a comparison with the original image considering Eq. (4). Figure 6 shows the result of such a regression, the four possible orientations and the value of the metric of similarity (4) derived from the regressed shape parameters of the four representations.

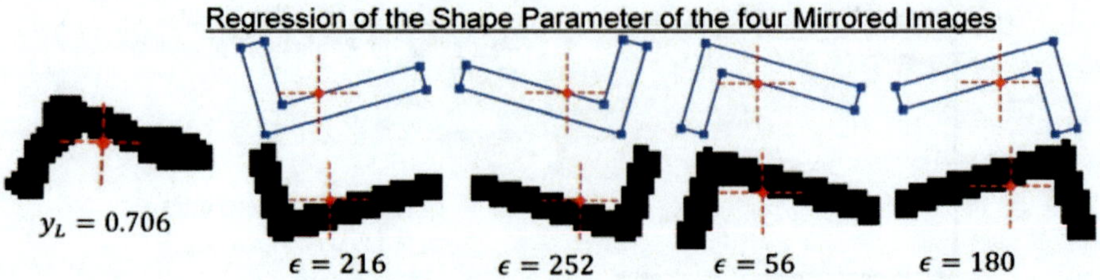

Fig. 6. Results of the regression and selection of the correct rotation. The third option leads to the smallest metric of similarity and thus defines the orientation to be selected

This regression can now be applied for different cross-sections and compared to the moments of area of the image. Figure 7 shows the result of such a classification and regression for an L-section and a Z-section. The shape parameter for these cross-sections appears appropriate compared to the image's visual appearance.

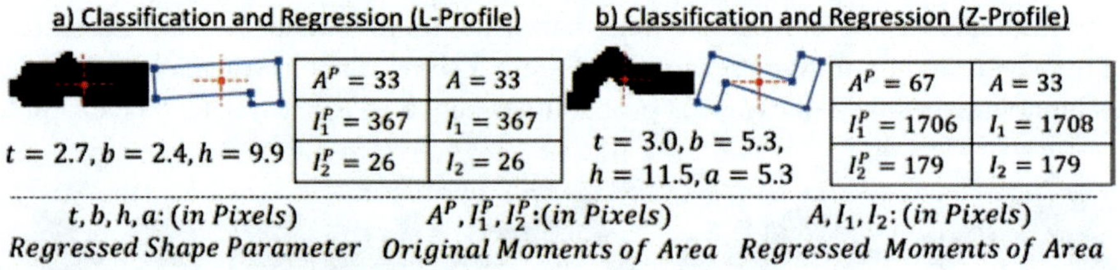

Fig. 7. Results of classification and regression and comparison of moments of area of the image cross-section and the regressed parameterized cross-section: a) regression and classification of an L-section ($y_L = 0.531$), b) regression and classification of a Z-section ($y_Z = 0.795$)

Comparing the cross-section properties of the images and the parametric representations shows only minor differences. This approach leads to a reasonable regression when the mechanical properties are considered. The proposed strategy is limited to the expression in (3) that must be minimized. If there are more than three parameters to regress, the system is underdetermined. In future works, the regression can be improved by also considering the image contours, resulting in a regression that considers mechanical properties and shape properties in combination. In addition, the data set for the classification task can be enriched by such cross-sections that come from industrial data samples instead of using only synthetic samples. Furthermore, the results can now be used for the automatic reconstruction of 3D geometries.

5 Summary

Parametrization of frame-like objects from polygon meshes or 3D images is a challenge in reverse engineering that requires recognition of the beamline and cross-sections. This article deals with the recognition of cross-section types and their shape parameter sets, which can be obtained by slicing a 3D image along an existing beamline. This involves successive classification of cross-section types via a neuronal network, followed by a regression of shape parameters by optimization of the objective function. Finally, the type of cross-section and its shape parameters can be derived from 2D images.

A significant focus of the work presented here is to exploit the properties of transformation invariant quantities of a cross-section similar to [23]. For example, the area of a triangle does not change with rotation or translation, so this quantity is considered invariant to rotation and translation. This property makes it possible to keep the dataset for training as small as possible since such a translation no longer needs to be learned. In this work, the data set could be halved by exploiting another invariance in contrast to [23].

In addition to the classification, the individual shape parameters must also be determined from the images. By defining an optimization task tailored to the respective class, the individual shape parameters were estimated taking into account the properties of the cross-section. The difference between the principal moments of area and the cross-sectional area of the image and the parametric description has to be minimized. However, the mere correspondence of their moments of area is not sufficient to identify the spatial orientation of the calculated shape parameters. In this work, an additional similarity comparison of the possible shape parameters with the original image is performed. All in all, this now leads to an automated classification of cross-section types and regression of the form parameters.

References

1. Hua Li and A. Yezzi, 'Vessels as 4D Curves: Global Minimal 4D Paths to Extract 3D Tubular Surfaces', in 2006 Conference on Computer Vision and Pattern Recognition Workshop (CVPRW'06), Jun. 2006, pp. 82–82. doi: https://doi.org/10.1109/CVPRW.2006.210.
2. K. Lidayová, H. Frimmel, C. Wang, E. Bengtsson, and Ö. Smedby, 'Fast vascular skeleton extraction algorithm', Pattern Recognit. Lett., vol. 76, pp. 67–75, Jun. 2016, doi: https://doi.org/10.1016/j.patrec.2015.06.024.
3. A. Nana, J.-C. Cuillière, and V. Francois, 'Automatic reconstruction of beam structures from 3D topology optimization results', Comput. Struct., vol. 189, pp. 62–82, Sep. 2017, doi: https://doi.org/10.1016/j.compstruc.2017.04.018.
4. G. Yin, X. Xiao, and F. Cirak, 'Topologically robust CAD model generation for structural optimisation', Comput. Methods Appl. Mech. Eng., vol. 369, p. 113102, Sep. 2020, doi: https://doi.org/10.1016/j.cma.2020.113102.
5. R. Bénière, G. Subsol, G. Gesquière, F. Le Breton, and W. Puech, 'A comprehensive process of reverse engineering from 3D meshes to CAD models', Comput.-Aided Des., vol. 45, no. 11, pp. 1382–1393, Nov. 2013, doi: https://doi.org/10.1016/j.cad.2013.06.004.

6. V. Vidal, C. Wolf, and F. Dupont, 'Mechanical Mesh Segmentation and Global 3D Shape Extraction', 2014.
7. B. Louhichi, G. N. Abenhaim, and A. S. Tahan, 'CAD/CAE integration: updating the CAD model after a FEM analysis', Int. J. Adv. Manuf. Technol., vol. 76, no. 1, pp. 391–400, Jan. 2015, doi: https://doi.org/10.1007/s00170-014-6248-y.
8. Y. M. Yoely, O. Amir, and I. Hanniel, 'Topology and shape optimization with explicit geometric constraints using a spline-based representation and a fixed grid', Procedia Manuf., vol. 21, pp. 189–196, Jan. 2018, doi: https://doi.org/10.1016/j.promfg.2018.02.110.
9. M. P. Bendsoe and O. Sigmund, Topology Optimization: Theory, Methods, and Applications, 2nd ed. Berlin Heidelberg: Springer-Verlag, 2004. doi: https://doi.org/10.1007/978-3-662-05086-6.
10. J. F. Aguilar Madeira, H. Rodrigues, and H. Pina, 'Multi-objective optimization of structures topology by genetic algorithms', Adv. Eng. Softw., vol. 36, no. 1, pp. 21–28, Jan. 2005, doi: https://doi.org/10.1016/j.advengsoft.2003.07.001.
11. J. Kresslein, P. Haghighi, J. Park, S. Ramnath, A. Sutradhar, and J. J. Shah, 'Automated cross-sectional shape recovery of 3D branching structures from point cloud', J. Comput. Des. Eng., vol. 5, no. 3, pp. 368–378, Jul. 2018, doi: https://doi.org/10.1016/j.jcde.2017.11.010.
12. A. Agathos, I. Pratikakis, S. Perantonis, N. Sapidis, and P. Azariadis, '3D Mesh Segmentation Methodologies for CAD applications', Comput.-Aided Des. Appl., vol. 4, no. 6, pp. 827–841, Jan. 2007, doi: https://doi.org/10.1080/16864360.2007.10738515.
13. D. Reniers and A. Telea, 'Part-type Segmentation of Articulated Voxel-Shapes using the Junction Rule', Comput. Graph. Forum, vol. 27, no. 7, Art. no. 7, 2008, doi: https://doi.org/10.1111/j.1467-8659.2008.01331.x.
14. M. Denk, K. Rother, and K. Paetzold, 'Fully Automated Subdivision Surface Parametrization for Topology Optimized Structures and Frame Structures using Euclidean Distance Transformation and Homotopic Thinning', in Proceedings of the Munich Symposium on Lightweight Design 2020, Berlin, Heidelberg, 2021, pp. 18–27. doi: https://doi.org/10.1007/978-3-662-63143-0_2.
15. M. Denk, K. Rother, and K. Paetzold, 'Subdivision Surfaces Mid-Surface Reconstruction of Topology Optimization Results and Thin-Walled Shapes using Surface Seletons', in Proceedings of the 21th International Conference on Engineering Design, Goethenburg, Aug. 2021, vol. 1, pp. 2771–2780. doi: https://doi.org/10.1017/pds.2021.538.
16. M. Denk, R. Klemens, and K. Paetzold, 'Beam Coloured Sketch-based 3D Shape Reconstruction', in Produktentwicklung und Design 2021, Dresden, in press, vol. 3.
17. A. Tagliasacchi, T. Delame, M. Spagnuolo, N. Amenta, and A. Telea, '3D Skeletons: A State-of-the-Art Report', Comput. Graph. Forum, vol. 35, no. 2, pp. 573–597, May 2016, doi: https://doi.org/10.1111/cgf.12865.
18. T. Stangl and S. Wartzack, 'Feature based interpretation and reconstruction of structural topology optimization results', in Proceedings of the 20th International Conference on Engineering Design (ICED15), Jul. 2015, p. Vol. 6, 235–245.
19. J. Mayer and S. Wartzack, 'Ermittlung eines Skelettierungsverfahrens zur Konvertierung von Topologieoptimierungsergebnissen', in Proceedings of the 31st Symposium Design for X (DFX2020), Bamberg, Sep. 2020, pp. 111–120. doi: https://doi.org/10.35199/dfx2020.12.
20. J.-C. Cuillière, V. François, and A. Nana, 'Automatic construction of structural CAD models from 3D topology optimization', Comput.-Aided Des. Appl., vol. 15, no. 1, pp. 107–121, Jan. 2018, doi: https://doi.org/10.1080/16864360.2017.1353726.
21. P.-S. Tang and K.-H. Chang, 'Integration of topology and shape optimization for design of structural components', Struct. Multidiscip. Optim., vol. 22, no. 1, pp. 65–82, Aug. 2001, doi: https://doi.org/10.1007/PL00013282.

22. A. Amroune, J.-C. Cuillière, and V. François, 'Automated Lofting-Based Reconstruction of CAD Models from 3D Topology Optimization Results', Comput.-Aided Des., vol. 145, p. 103183, Apr. 2022, doi: https://doi.org/10.1016/j.cad.2021.103183.
23. M. Denk, K. Rother, T. Höfer, J. Mehlstäubl, and K. Paetzold, 'Euclidian Distance Transformation, Main Axis Rotation and Noisy Dilitation supported Cross-Section Classification with Convolutional Neuronal Networks', presented at the International Conference on Engineering Design (ICED21), Goethenburg, in press.
24. R. T. Marler and J. S. Arora, 'The weighted sum method for multi-objective optimization: new insights', Struct. Multidiscip. Optim., vol. 41, no. 6, pp. 853–862, Jun. 2010, doi: https://doi.org/10.1007/s00158-009-0460-7.

Clustering Topologically-Optimized Designs Based on Structural Deformation

Ernest Hutapea[1], Nivesh Dommaraju[1(✉)], Mariusz Bujny[2], and Fabian Duddeck[1]

[1] Technical University of Munich, Arcisstraße 21, 80333 Munich, Germany
ernest_hutapea@yahoo.com, nivesh.dommaraju@tum.de, duddeck@tum.de
[2] Honda Research Institute Europe GmbH, Carl Legien Str. 30, 63073 Offenbach/Main, Germany
mariusz.bujny@honda-ri.de

Abstract. Topology optimization can be used to generate a large set of lightweight structural solutions either by changing the constraints or the weights for different objectives in multi-objective optimization. Engineers must analyze and review the designs to select solutions according to their preference towards objectives such as structural compliance and crash performance. However, the sheer number of solutions challenge the engineers' decision-making process. An automated way of summarizing solutions is to cluster groups of similar designs based on a suitable metric. For example, with the Euclidean metric in the objective vector, design groups with similar performance can be identified and only the representative designs from the different clusters may be analyzed. Since the deformation behavior of a structure is an important design feature, in this work, we investigate the use of manifold learning algorithms to identify and group similar designs using the nodal displacement data. The proposed approach can process the volumetric deformation of geometries with completely different topologies. In this study, we couple the manifold learning techniques, t-distributed Stochastic Neighbor Embedding (t-SNE) and Uniform Manifold Approximation and Projection (UMAP), with the clustering algorithms, k-means and Ordering Points To Identify the Clustering Structure (OPTICS), to identify the representative deformation modes. Using Gaussian Random Fields (GRF) to create artificial displacement fields, we generate a labeled dataset with different modes, which enabled us to evaluate our method using classification accuracy, precision, recall, and F1-score. Finally, using our approach, we successfully distinguished between similar and non-similar designs in the results from topology optimization.

Keywords: Deformation behavior · Topology optimization · Manifold learning · Gaussian random field · Classification metrics

J. Rieser et al. (Eds.): *Proceedings of the Munich Symposium on Lightweight Design 2021*, Proceedings, pp. 104–114, 2023. https://doi.org/10.1007/978-3-662-65216-9_10

1 Introduction

Topology Optimization (TO) [1] is a mathematical method that optimizes material layout within a given design space for a given set of loads, boundary conditions, and constraints, with the goal of maximizing the performance of a structure. Thanks to the progress in the manufacturing technologies, like 3D-printing, as well as the latest research developments in the TO field, these methods are increasingly used across different disciplines, including civil engineering [2] and vehicle crashworthiness [3–9]. When conflicting objectives such as crash energy absorption and structural compliance are used, a multi-objective TO approach creates a Pareto front of optima, from which engineers need to select solutions using additional criteria such as appearance [10].

A large set of solutions might impede the engineers' decision-making process. To mitigate this problem, Hagg et al. [11] cluster similar solutions and obtain a representative solution from each group. They use machine learning methods, such as Principal Component Analysis (PCA) [12], autoencoder [13], t-distributed Stochastic Neighbor Embedding (t-SNE) [14], kernel-PCA [15], and Isomap [16] to map the set of solutions to a similarity space and cluster similar solutions into different classes. Similarly, Dommaraju et al. [10] use PCA to develop a metric for geometrical differences and identify prototypical designs in the Pareto front. In a separate research paper [17], they use an autoencoder, a non-linear dimensionality reduction method, to identify geometrical prototypes.

Since the deformation behavior of a structure is important, we investigate methods to cluster based on deformation modes. Garcke et al. [18] and Sible et al. [19] demonstrate the use of non-linear dimensionality reduction methods on the displacement fields of a set of structures to find designs with similar deformation modes, including complex deformations from crash simulations. However, their approaches might not be suitable for TO designs since they assume isometric deformation. Furthermore, we want to analyze TO designs with very different topologies. Another interesting approach, proposed by Diez et al. [20], combines decision tree and rule mining to identify different deformation behaviors in crash simulation results but it requires labeled datasets.

In this study, we propose the use of manifold learning methods, t-SNE and Uniform Manifold Approximation and Projection (UMAP) [21], to perform dimensionality reduction on the nodal displacement data. The resultant data can then be processed by clustering algorithms such as k-means [22] and OPTICS (Ordering Points To Identify the Clustering Structure) [23] to identify structures with similar deformation behavior. Dimensionality reduction methods require that the nodal displacement data is consistent across the structures. So, we retain the elements that are allotted very small densities by the TO method.

To evaluate the performance of the proposed methods, we use two types of datasets. The first dataset is created by applying artificial displacement fields on cube-shaped structures to obtain different deformation modes. We can then investigate if our proposed approach correctly identifies the different modes. The second dataset contains TO results. While the first labeled dataset allows us to evaluate our approach using classification metrics, the second dataset allows only visual evaluation.

This paper is structured as follows. In Sect. 2, we briefly discuss the proposed workflow of our method. In Sect. 3, we discuss the two types of datasets and how to generate them while the results and conclusions are presented in Sects. 4 and 5, respectively.

2 Method

Fig. 1 shows the proposed workflow to find the deformation modes where N is the number of structures in the set, M is the dimension of the nodal displacement field, and P is the number of dimensions in the similarity space produced by the manifold learning algorithm.

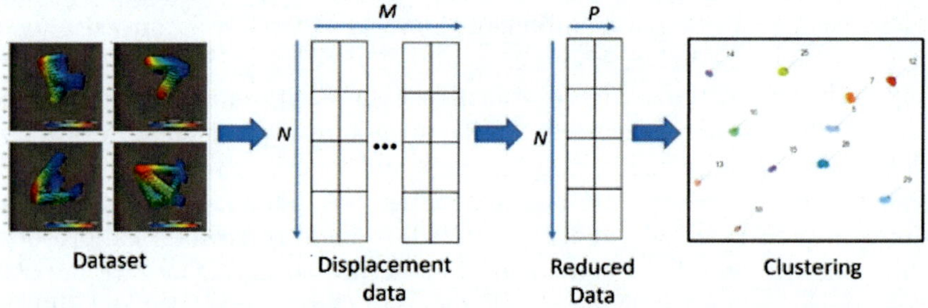

Fig. 1. General workflow to obtain design clusters. Displacement data are preprocessed into a matrix of size $N \times M$, then a manifold learning algorithm reduces the size of the matrix to $N \times P$. Finally, a clustering algorithm tries to find clusters and obtain representative designs from each cluster

Instead of removing the low-density elements, data of all the nodes in the cube is used as input for the dimensionality reduction method to ensure consistency of data across all designs. Therefore, with our approach, we can compare the deformation behavior of topologically different structures, as shown in Sect. 4.

To reduce the dimensionality of the nodal displacement field, we use the manifold learning methods, UMAP, and t-SNE, which non-linearly reduce the data. We consider two different clustering algorithms: k-means [22], a popular inexpensive clustering algorithm, and OPTICS [23], a density-based clustering algorithm. Silhouette method [24] is used to set the appropriate number of clusters for k-means algorithm.

3 Dataset Generation

In this section, we discuss the datasets used to test our approach, namely the artificial and topology optimization datasets. The initial finite element mesh that we use for both types of dataset is shown in Fig. 2.

3.1 Labeled Test Dataset

3.1.1 Gaussian Random Field

To generate a labeled test dataset of deformed structures, we apply displacement fields generated by Gaussian Random Field (GRF) [25] on cube-shaped structures. GRF determines the displacement field using the mean and the covariance function. Given the displacements of specified nodes, the mean function determines the most probable positions of the unspecified nodes while the covariance function incorporates the uncertainty associated with the field. Using GRF, we can easily specify the deformation mode and generate the corresponding deformed structures much more quickly compared to finite element simulations.

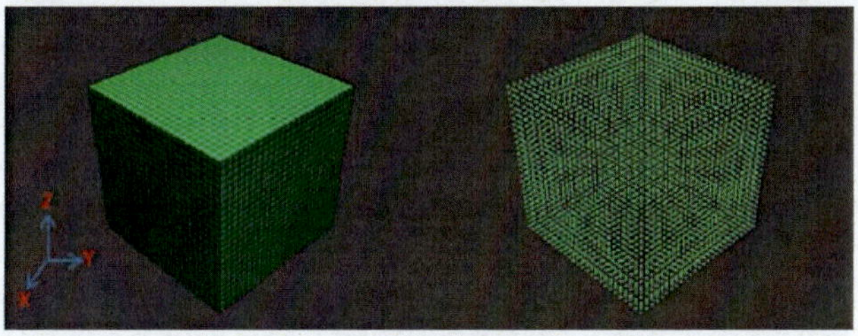

Fig. 2. Finite element model in a cube domain, each side is 1 m in length, with 15,625 elements. The material properties are mass density = 7.83 10^{-9} ton mm^{-3}, Young's modulus = 207 GPa, and Poisson's ratio = 0.33

3.1.2 Dataset Generation

To generate the displacement fields seen in mechanical structures, we compose the displacement field using the main field and the noise field. We also add randomness to the selected node positions, displacement magnitudes, and displacement directions to increase the variation across designs and to challenge our approach.

To apply the main field on an undeformed cube structure, we randomly select one or more nodes on the surface of the cube. A few examples are shown in the top row of Fig. 3. The selected two nodes act as specified nodes while the remaining nodes act as unspecified nodes. After that, we can calculate the mean displacement of the structures using the mean of GRF. The resulting set of structures are shown in the bottom row of Fig. 3.

To generate a noise field, it is possible to use the covariance matrix of a GRF. However, due to the large number of nodes in our cube structure, it is too expensive to create the covariance matrix. Therefore, we use a different approach to create the noise field. We randomly select 100 nodes on the cubic surface, which are then assigned displacement vectors with random magnitudes and directions. To deform the structure, we can compute the mean node positions using GRF. The resulting cube structure with noise field is shown in Fig. 4. By adding different noise fields to the main field, multiple deformations of a single mode can be obtained. The resulting structures are shown in Fig. 5.

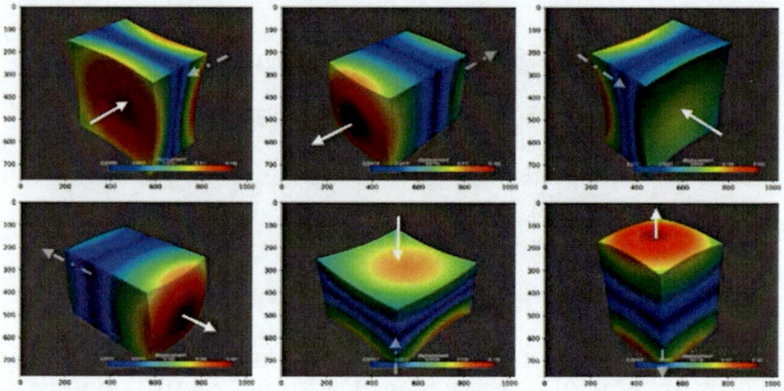

Fig. 3. Examples of our labeled test dataset. We apply displacement vectors on two opposing faces of the cube

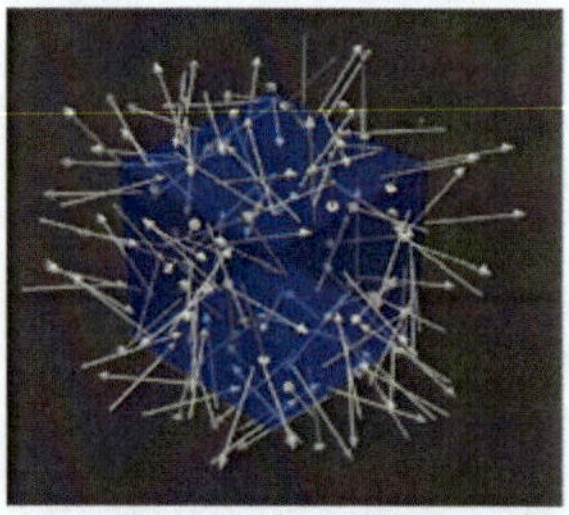

(a) Select random nodes on the surface

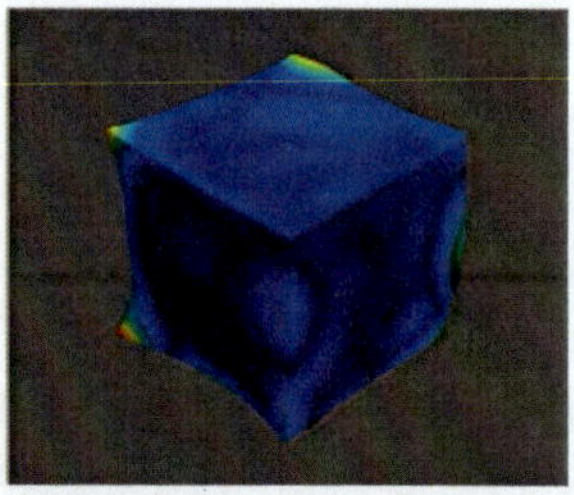

(b) Deformed cube model

Fig. 4. Deforming model with noise field

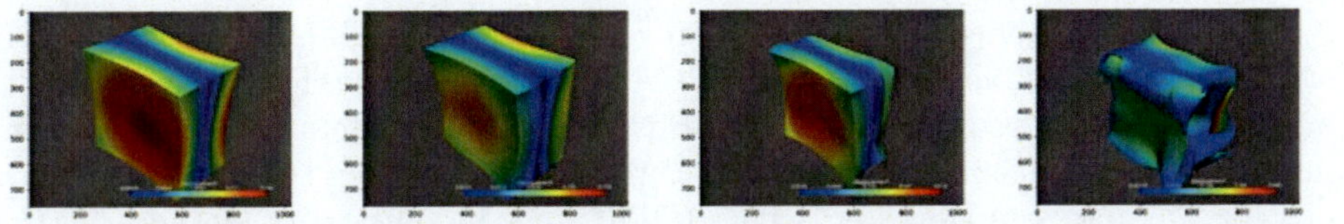

(a) Noise level=0% (b) Noise level=20% (c) Noise level=50% (d) Noise level=100%

Fig. 5. Models deformed using different noise levels

To make the separation of the deformation modes more challenging, the amount of noise level for each mode can be varied. Noise level is defined as the ratio between the maximum displacement of the noise field and the maximum displacement of the main field. Deformed models with different noise levels are shown in Fig. 5 where the maximum displacement of the main field is 0.2 m. For visualization purpose, the displacement is scaled up by a factor of two. For higher noise levels, the deformations become less realistic. So a noise level of 50% is chosen.

In total, there are 5 sets of node selections. Each set of node selections produces 6 different deformation modes, with 20 structures for each mode. Therefore, we have 600 structures in this dataset with 30 deformation modes.

3.2 Topology Optimization Design Dataset

The dataset of TO designs used in this paper are the results of Solid Isotropic Material with Penalization (SIMP) TO using the optimality criteria (OC) method [1]. We use the following parameters in SIMP: move limit is 0.1, filter radius is 0.1, volume fraction is 0.3, and penalization factor is 3. For each optimization, one of the faces is fixed and the cubic structure is optimized for two arbitrarily applied loads. In total, there are 100 topologically different structures in this dataset. Several structures in the TO design set can be seen in Figs. 9 and 10.

4 Results and Analysis

4.1 Results on Labeled Test Dataset

In this subsection, we show the results of applying our approach on the labeled test dataset with a noise level of 50%. In Fig. 6, we show the plots of using UMAP with k-means (Fig. 6) and OPTICS (Fig. 6). The x and y axes are the reduced dimensions, created by the UMAP algorithm. The clusters, i.e., the deformation modes, obtained by the clustering algorithms, are represented in different colors and are assigned different numbers, as pointed at by the arrows. As seen in Fig. 6, both k-means and OPTICS could easily cluster the structures in our dataset, although k-means does not perform well when the clusters are located very close to each other.

We investigate the clustering accuracy obtained using UMAP to other dimensionality reduction methods: PCA, k-PCA, and t-SNE. For reference, results with no dimensionality reduction are also used. With k-means clustering, there is no noticeable difference in classification accuracy (0.6) between the reduction methods for this dataset. However, with OPTICS as the clustering method, UMAP outperforms other methods (Fig. 7).

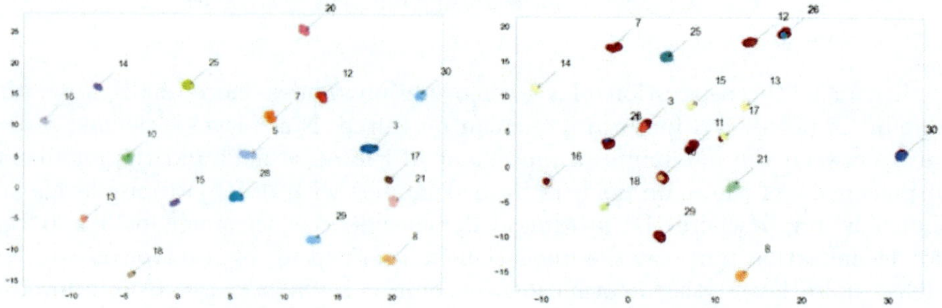

(a) Clusters labeled by k-means

(b) Clusters labeled by OPTICS

Fig. 6. Identified clusters in the artificial dataset using UMAP. In both the plots, each design is shown as a point whose x- and y-coordinates are the corresponding UMAP components, while each cluster of points is annotated and shown with a distinct color

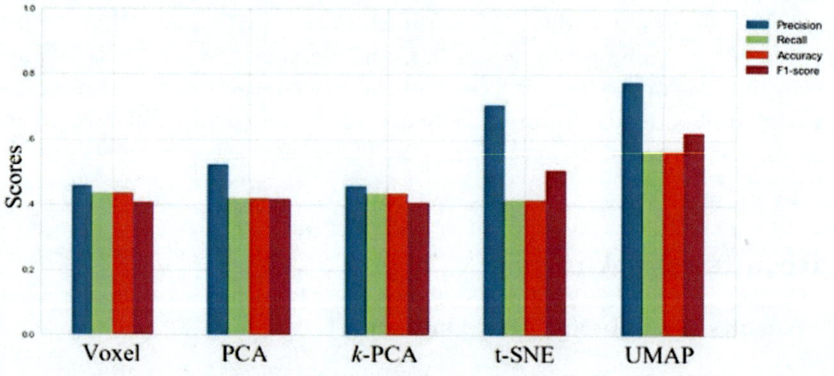

Fig. 7. Comparison of different dimensionality reduction methods based on precision, recall, accuracy, and F1-score. OPTICS is used for clustering designs

4.2 Results on to Design Dataset

In this subsection, we apply UMAP with k-means clustering on the TO design dataset. Using the silhouette method, we select 18 as the proper number of clusters. To show that our proposed cluster designs with similar deformation behavior, we visualize four structures from two clusters numbered 11 and 2 (Fig. 8).

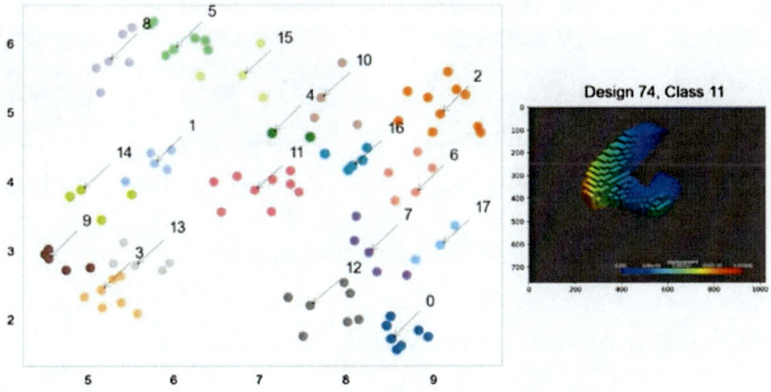

Fig. 8. TO designs mapped to 2-D using UMAP components

Figure 9 shows four structures in cluster 11 which have similar circular deformation pattern in the middle, despite having different underlying topologies. Figure 10 shows another example cluster. Here, the structures have pronounced deformation in the fore-most corner.

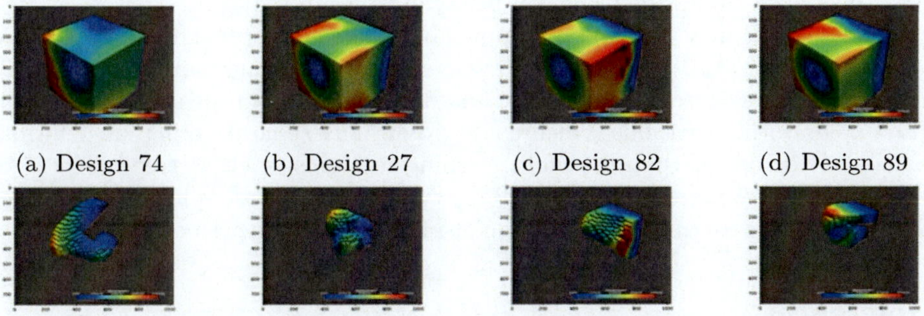

(a) Design 74 (b) Design 27 (c) Design 82 (d) Design 89

(e) Design 74 (f) Design 27 (g) Design 82 (h) Design 89

Fig. 9. Different designs from class 11

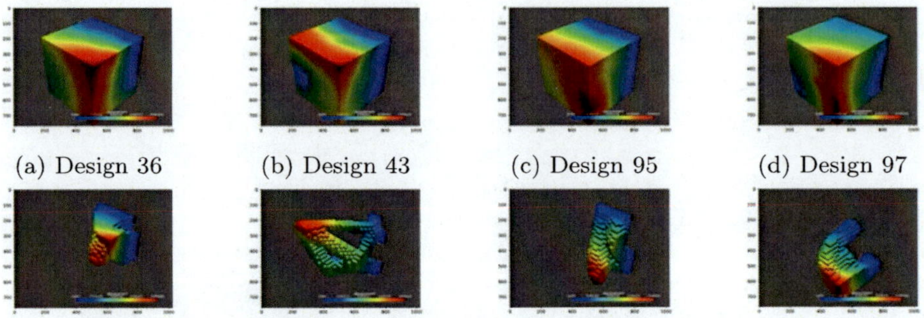

(a) Design 36 (b) Design 43 (c) Design 95 (d) Design 97

(e) Design 36 (f) Design 43 (g) Design 95 (h) Design 97

Fig. 10. Different designs from class 1

5 Conclusion and Future Work

In this paper, we address the problem of clustering a large set of topologically different TO designs, based on the deformation behavior. We use manifold learning methods to perform dimensionality reduction and then use clustering algorithms to cluster designs with similar deformation behaviors. For a consistent node-to-node comparison, we retain the low-density elements of the TO designs. The main methods that we use for our work are t-SNE and UMAP for dimensionality reduction, and k-means and OPTICS to perform clustering.

To test our approach, we generated two datasets: a labeled dataset generated with GRF and a TO design dataset. With our approach, we successfully identified structures with similar deformation behavior on both datasets.

Future work may explore the applicability of our approach using different mechanical features, methods, or datasets. Interesting mechanical features to explore are stress, strain, geometries, and displacement as time series. We might also explore other methods which could potentially separate different deformation behavior better than manifold learning methods, such as tree-based algorithms or neural-networks. If available, we could test our approach on larger datasets with more diverse TO designs to further test its performance.

References

1. Bendsøe M, Sigmund O (2004) Topology Optimization: Theory, methods and applications. Springer, https://doi.org/10.1007/978-3-662-05086-6
2. Stromberg L, Beghini A, Baker W, Paulino G (2012) Topology optimization for braced frames: Combining continuum and beam/column elements. Engineering Structures 37:106-124, https://doi.org/10.1016/j.engstruct.2011.12.034
3. Patel N, Kang BS, Renaud J, Tovar A (2009) Crashworthiness design using topology optimization. Journal of Mechanical Design 131(6), https://doi.org/10.1115/1.3116256

4. Ortmann C, Schumacher A (2013) Graph and heuristic based topology optimization of crash loaded structures. Structural and Multidisciplinary Optimization 47, https://doi.org/10.1007/s00158-012-0872-7
5. Duddeck F, Hunkeler S, Lozano P, Wehrle E, Zeng D (2016) Topology optimization for crashworthiness of thin-walled structures under axial impact using hybrid cellular automata. Structural and Multidisciplinary Optimization 54, https://doi.org/10.1007/s00158-016-1445-y
6. Bujny M, Aulig N, Olhofer M, Duddeck F (2018) Identification of optimal topologies for crashworthiness with the evolutionary level set method. International Journal of Crashworthiness 23:1–22, https://doi.org/10.1080/13588265.2017.1331493
7. Raponi E, Bujny M, Olhofer M, Aulig N, Boria S, Duddeck F (2019) Kriging-assisted topology optimization of crash structures. Computer Methods in Applied Mechanics and Engineering 348, https://doi.org/10.1016/j.cma.2019.02.002
8. Yousaf MS, Bujny M, Zurbrugg N, Detwiler D, Duddeck F (2021) Similarity control in topology optimization under static and crash loading scenarios. Engineering Optimization 53(9):1523–1538, https://doi.org/10.1080/0305215X.2020.1806257
9. Bujny M, Olhofer M, Aulig N, Duddeck F (2021) Topology optimization of 3d-printed joints under crash loads using evolutionary algorithms. Structural and Multidisciplinary Optimization https://doi.org/10.1007/s00158-021-03053-4
10. Dommaraju N, Bujny M, Menzel S, Olhofer M, Duddeck F (2020) Simultaneous Exploration of Geometric Features and Performance in Design Optimization. In: 16th International LS-DYNA Conference, p 12
11. Hagg A, Asteroth A, Bäck T (2018) Prototype Discovery Using Quality-Diversity. In: Int. Conf. on Parallel Problem Solving from Nature, Springer, pp 500–511, https://doi.org/10.1007/978-3-319-99253-2_40
12. Pearson K (1901) LIII. On lines and planes of closest fit to systems of points in space. https://doi.org/10.1080/14786440109462720
13. Hinton GE, Salakhutdinov RR (2006) Reducing the Dimensionality of Data with Neural Networks. Science 313(5786):504–507, https://doi.org/10.1126/science.1127647
14. van der Maaten L, Hinton G (2008) Visualizing data using t-sne. Journal of Machine Learning Research 9:2579–2605
15. Schölkopf B, Smola A, Müller KR (1997) Kernel principal component analysis. In: International Conference on Artificial Neural Networks, Springer Berlin Heidelberg, pp 583–588
16. Tenenbaum J, Silva V, Langford J (2001) A global geometric framework for nonlinear dimensionality reduction. Science (New York, NY) 290:2319–23, https://doi.org/10.1126/science.290.5500.2319
17. Dommaraju N, Bujny M, Menzel S, Olhofer M, Duddeck F (2019) Identifying Topological Prototypes using Deep Point Cloud Autoencoder Networks. In: 2019 Int. Conf. on Data Mining Workshops (ICDMW), IEEE, pp 761–768, https://doi.org/10.1109/icdmw.2019.00113
18. Garcke J, Iza-Teran R (2017) Machine Learning Approaches for Data from Car Crashes and Numerical Car Crash Simulations. In: Int. Conf. Simulation Process & Data Management (SPDM)
19. Sible S, Iza-Teran R, Garcke J, Aulig N, Wollstadt P (2020) A Compact Spectral Descriptor for Shape Deformations. In: European Conference on Artificial Intelligence, IOS Press, vol 325, pp 1930–1937, https://doi.org/10.3233/FAIA200311
20. Diez C, Kunze P, Toewe D, Wieser C, Harzheim L, Schumacher A (2018) Big-Data Based Rule-Finding for Analysis of Crash Simulations. In: Advances in Structural

and Multidisciplinary Optimization, pp 396–410, https://doi.org/10.1007/978-3-319-67988-4_31

21. McInnes L, Healy J, Saul N, Großberger L (2018) UMAP: Uniform Manifold Approximation and Projection. Journal of Open Source Software 3(29):861, https://doi.org/10.21105/joss.00861
22. MacQueen J, et al (1967) Some methods for classification and analysis of multivariate observations. In: Proceedings of the fifth Berkeley symposium on mathematical statistics and probability, Oakland, CA, USA, vol 1, pp 281–297
23. Ankerst M, Breunig MM, Kriegel HP, Sander J (1999) OPTICS: Ordering Points to Identify the Clustering Structure. SIGMOD Record (ACM Special Interest Group on Management of Data) 28(2):49-60, https://doi.org/10.1145/304181.304187
24. Rousseeuw PJ (1987) Silhouettes: a graphical aid to the interpretation and validation of cluster analysis. Journal of Computational and Applied Mathematics 20:53–65
25. Hristopulos DT (2020) Gaussian Random Fields, Springer Netherlands, Dordrecht, pp 245–307. https://doi.org/10.1007/978-94-024-1918-4_6

Optimization of Fused Filament Fabricated Infill Patterns for Sandwich Structures in a Three-Point Bending Test

Tobias Rosnitschek(✉), Annika Gläseke, Florian Hüter, Bettina Alber-Laukant, and Stephan Tremmel

Engineering Design and CAD, University of Bayreuth, 96447 Bayreuth, Germany
{tobias.rosnitschek,Annika.Glaeseke,florian.hueter, bettina.alber,stephan.tremmel}@uni-bayreuth.de

Abstract. Additive manufacturing is taking on an essential role in lightweight design engineering. To realize significant weight reductions and manufacturing time savings, additively manufactured components often consist of a shell and an inner infill structure. By now, infill structures are mainly simple geometrical patterns without any link to the load cases of the components.

In this article, we investigate the possibility of using topology optimization for more dedicated infill structures. Therefore, we present a workflow to create the topology-optimized shape of a unit cell, patterned and extruded as an infill structure. The workflow is used to optimize various three-point bending specimens, which are manufactured by fused filament fabrication. The specimens are compared in experiments and finite element simulations utilizing a lightweight number. Though open challenges exist in further automation and consideration of other load cases, shape orientation and additive manufacturing processes, this contribution represents a basic approach to employing topology optimization to optimize infill structures.

Keywords: Additive Manufacturing · Finite Element Analysis · Infill Structures · Sandwich Components · Topology Optimization

1 Infill Structures in Additive Manufacturing

In subtractive manufacturing, lightweight applications cause higher production costs since more material has to be removed, corresponding to more machining operations. Also, it is often hardly possible to realize the complexity of topology optimized parts for lightweight applications with conventional manufacturing processes. Adversely, in additive manufacturing (AM), lightweight applications are especially desired. The freedom of design in AM widens the possibilities for adding geometrical complexity at no extra costs [1, 2].

J. Rieser et al. (Eds.): *Proceedings of the Munich Symposium on Lightweight Design 2021*, Proceedings, pp. 115–123, 2023. https://doi.org/10.1007/978-3-662-65216-9_11

Furthermore, since machine operations in AM are adding material, the less material is needed for the part, the faster and cheaper the manufacturing process is. Hence, lightweight engineering is explicitly effective in exploiting the full potential of additive manufacturing. To lift this potential, the design for AM (DfAM) methodology offers multiple levers, such as, for instance, structural, or in particular, topology optimization and infill structures [3]. Topology optimization (TO) has gained a certain maturity in recent years. It is one of the leading tools for structural optimization in computer-aided engineering [4] and a well-established research field with current vital areas such as automated redesign of topology optimized structures or integrating design for AM aspects [4, 5]. Today, most approaches still assume solid parts and do not consider structure-filled components as they are mostly used in AM. One example for dealing explicitly with the anisotropic features of AM processes in TO can be found in [6]. This example considers the combination of shell and infill but does not interact with the infill structure. Nevertheless, the consideration of the infill pattern shows high potential for changing the part's effective properties, as [7] shows exemplary that the modification of the infill degree can vary the effective Young's modulus of a part significantly and, therefore, easily allows parts with graded properties. The main purpose of infill structures is to reduce material consumption and printing time. Hence, they are selected primarily under production engineering aspects and not from a mechanical point of view. Gopsill et al. [8] address this point and modify the infill structure globally by adding load-path-dependent additional struts to some infill cells. However, for this approach, a finite element analysis, or alternatively, TO of the part has to be conducted beforehand to obtain the load-paths.

In this article, we also consider infill patterns from a mechanical point of view, but instead of changing the shape of some infill cells, our approach is to define infill cells adapted to the part's primary load case. This will enable standardized infill patterns for primary load cases without conducting preliminary numerical analysis of the system.

2 Load-Case-Oriented Infill Patterns

Most infill patterns for AM are simple geometrical patterns, such as hexagonal or triangular honeycombs. Their primary purpose is production orientated and has the objective to reduce the needed material and printing time. Conventionally an infill pattern is made by replacing a solid cell with a 2.5-dimensional lattice structure patterned over the whole cross-section and extruded in the part's stacking direction, as illustrated in Fig. 1.

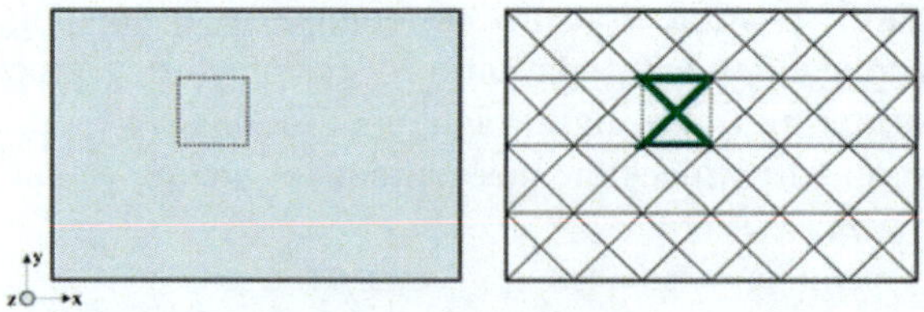

Fig. 1. Creating a 2.5-dimensional lattice infill pattern for an arbitrary cross-section. The unit cell (green) replaces the solid cell and is used as infill pattern, which is extruded in stacking direction (z)

Our approach for load-case-oriented infill patterns is based on the idea of designing infill patterns adapted from the part's principal load, for instance, tensile-, bending- or shear-load.

Therefore, we conduct a 2-dimensional TO (2D-TO) for minimal compliance with the infill cell and its associated loading, as depicted in Fig. 2. This allows us to provide standardized infill patterns which are dedicated to basic loading conditions. In this way, specific parts or areas of parts can easily be adapted to provide a more load-case-oriented inner structure.

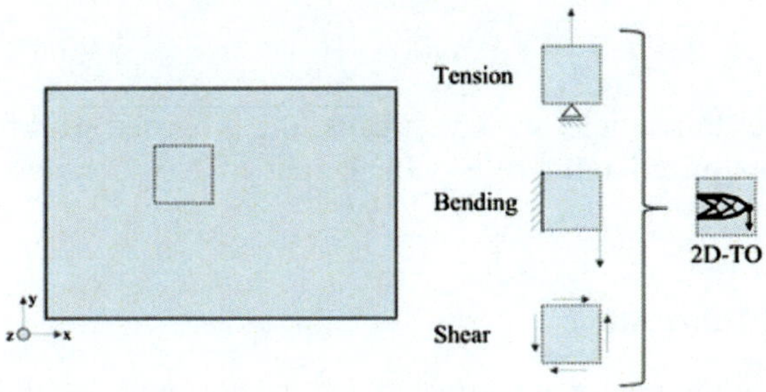

Fig. 2. Examples of basic loading conditions, which used in 2D-TO to obtain optimized infill patterns

Hence, our workflow for creating load-case-oriented infill patterns consists of the following four steps:

1. Define the shape of the infill cell
2. Conduct a 2D-TO for the primary load case on the infill cell
3. Convert the 2D-TO result to a 2D-geometry infill cell
4. Use the optimized cell as an infill pattern for the part

For the 2D-TO and the conversion to a 2D-geometry infill cell, we slightly adapted the python code from [9] to solve a minimum compliance problem with a volume constraint which can be written as.

$$\min \boldsymbol{C} = \min \boldsymbol{u}^T \boldsymbol{K} \boldsymbol{u}$$

$$\text{s.t.} : v \leq V_T \ (1) \tag{1}$$

where $\boldsymbol{C}$ denotes the Compliance, $\boldsymbol{u}$ the displacement vector, and $\boldsymbol{K}$ the stiffness matrix. The constraint is written as the relative volume v, which has to be equal or smaller than the target relative Volume V_T. The obtained result is then converted to a point map and stored as a CSV file. Followingly, we use the software nTopology [10] to pattern the optimized 2D infill cell across the part and create the final infill structure. The result is exported as an STL file and transferred to the slicing software.

Figure 3 shows exemplary results of optimizing a quadratic infill cell for bending load from the greyscale image to the extruded infill cell.

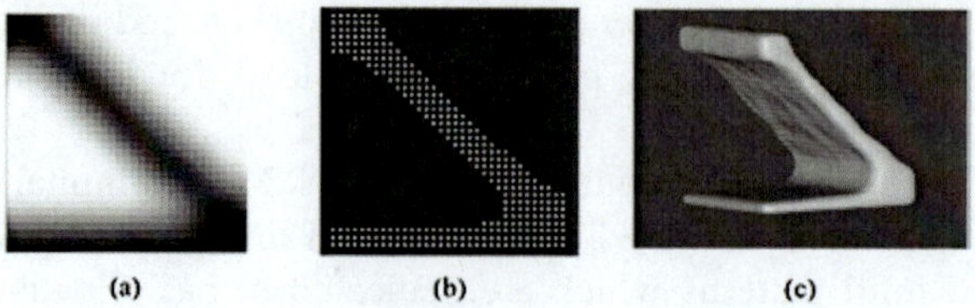

Fig. 3. Exemplary results for creating a bending-optimized infill cell based on a quadratic shape. **a**) greyscale image after 2D-TO; **b**) point map after post processing; **c**) extruded final infill cell profile

In the scope of this article, we investigated the proposed workflow by the example of bending-optimized infill cells with height to length ratios of 1:1, 1:2, 1:5, respectively.

2.1 Design and Fabrication

Our proposed workflow was investigated by the example of three-point bending beams with dimensions length = 200 mm, width = 40 mm, and height = 30 mm. The beams consisted of a top and bottom layer of 2 mm, the remaining core was replaced by the respective infill pattern. For the optimization the solid isotropic material interpolation with penalization (SIMP) approach was used, there the element's Young's Modulus E is described the initial Young's Modulus E_0, the element's relative density and a penalty factor p.

$$E(\rho_i) = {\rho_i}^p E_0. \tag{2}$$

In all optimizations, we used a target volume of 0.25, a penalty factor of 3 and considered all elements with a design variable greater than 0.4 for the post processing.

All parts in the following were printed via the Fused Filament Fabrication Process (FFF) with a layer height of 0.2 mm, using the Markforged MarkTwo and the nylon based Markforged Onyx filament, the slicing was done with the Markforged Eiger software [11, 12]. The settings for infill optimization and the resulting volumes and masses are summarized in Table 1 and their shapes together with the TO boundary conditions are depicted in Fig. 4.

Table 1. Settings and resulting volumes and masses for the printed specimens. The volume was calculated by Eiger, while the mass was measured after fabricating the specimens

Name	Aspect ratio	Cell size (x, y) in mm	Target volume	Volume in cm^3	Mass in g
B1	1:1	10, 10	0.25	70.79	70.7
B2	1:2	10, 20		133.89	145.0
B3	1:5	10, 50		105.36	115.0

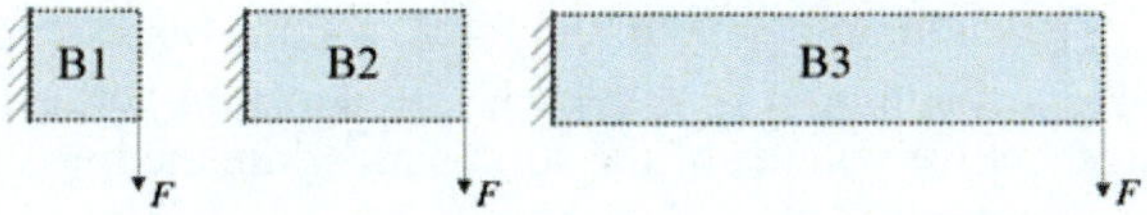

Fig. 4. The shape and TO boundary conditions for the considered specimens

2.2 Experimental Procedure

A three-point bending test (Fig. 5.) was performed with three specimens for each infill pattern configuration. The procedure was aligned to ASTM C393/393 M with a support span of 15 mm. As stopping criteria, a displacement of 10 mm, respectively core crushing were chosen.

Fig. 5. Picture of the used experimental setup

2.3 Evaluation

We analyzed the results by different metrics. The maximum force F_{max} is used to compare the overall performance of the infill patterns. Also, for a quantitative evaluation, the lightweight number M is useful, which can be written the following according to [13, 14]:

$$M = \frac{F}{mg} \tag{3}$$

where F is a defined force, m is the part's mass, and g is the gravitational acceleration. We use M in two different variations. Firstly, we compared the specimens in the linear elastic regime and evaluated M_1 for F_1 at a displacement of 1 mm, and second, we evaluated M_{max} for the maximum forces F_{max}.

Furthermore, the specimens were compared by the effective density-related-stiffness S_{eff}, which was calculated by:

$$S_{eff} = \frac{F_{max}}{d_{F_{max}}} \frac{1}{\rho_{eff}} \tag{4}$$

with

$$\rho_{eff} = \frac{V_{init}}{m} \tag{5}$$

where F_{max} and d_{Fmax} denoting the maximum force, and the displacement at maximum force, respectively. As this metric is defined for effective properties, the volume V_{init} describes the representative volume of the solid part, while the mass m belongs to the actual printed plastic volume.

2.4 Finite Element Analysis

For the finite element analysis (FEA) of the three-point bending test, we applied the Hencky model [15], which can be used for geometric nonlinear problems with linear elastic material behavior. We considered the symmetry of the specimens and used quarter models for the FEA to reduce the computational effort. Further, all models were meshed with quadratic tetrahedrons, and a three-point bending test was simulated until a displacement of 10 mm. We conducted a compression test to obtain the Young's modulus, resulting in a value of 0.7 GPa. However, it is notable that this result differed significantly from the value of 2.5 GP given in the data-sheet [12]. We attribute this difference to manufacturing-related defects like pores which arise from the use of different printers and print settings. Accordingly, for all simulations, we set the Young's modulus to 0.7 GPa and the Poisson's ratio to 0.35, based on the literature values for nylon [16]. All simulations were conducted with Z88, an FEA freeware developed by Engineering Design and CAD at the University of Bayreuth [17].

3 Results and Discussion

Based on the cell sizes and TO settings, we obtained the following three cells depicted in Fig. 6, outlined in yellow, which were used to create the infill patterns of the specimens. As a starting point for the pattern the geometric center point was chosen.

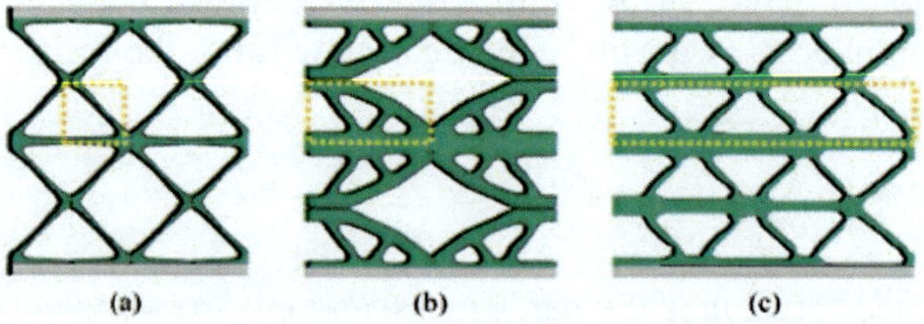

Fig. 6. Optimized infill patterns after 2D-TO; **a)** B1, **b)** B2 and **c)** B3

The optimized cell of B1 is very similar to the commonly used triangular infill pattern, and B3 is basically a combination of multiple B1 cells. The result of B2, meanwhile, looks quite different and is very close to the well-known TO-design proposal for minimum compliance cantilever problems [18, 19]. Since B2 has the highest mass, it was also assumed to show the best experimental performance, which could be confirmed during testing. As the test results in Fig. 7 show, the maximum force of B2 is more than twice the maximum force of B3 and nearly four times the maximum force of B1. The measurements show the expected behavior of a linear regime followed by a plateau, which is caused by core-crushing. Since the strut diameters of B1 and B3 are thinner as of B2, it is obvious that their measurements show a more distinctive plateau.

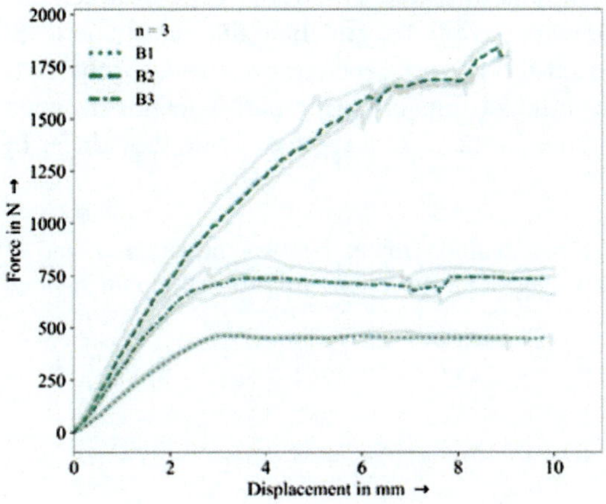

Fig. 7. Testing results of the optimized infill patterns. The grey shaded lines are the actual measurements, while the green lines represent mean curves computed by rigid regression

By comparing the FEA results with the experiments in Fig. 8, the simulations capture the trend but slightly underestimate the samples' stiffness from the experimental results. Hence, with a more detailed study on the material constants, the Hencky model is a good choice for the simulation of structure filled additively manufactured parts.

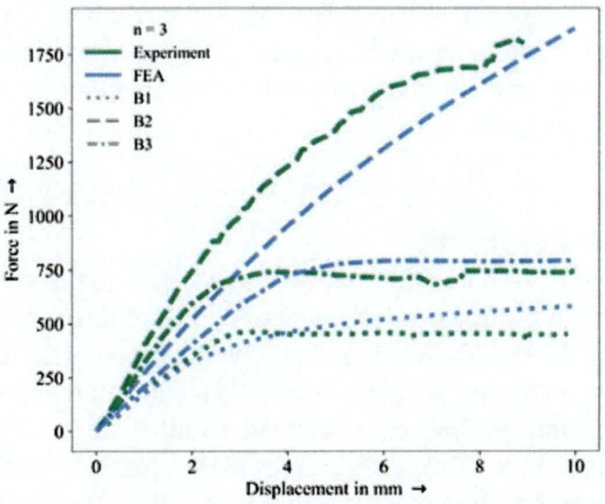

Fig. 8. Comparison of experimental (green) and FEA (blue) results. The qualitative trend could be reproduced but the simulations underestimate the stiffness of the specimens slightly

Furthermore, the comparison showed, that a Young's modulus of 0.7 GPa was much more appropriate than the data-sheet value of 2.5 GPa.

Evaluating the metrics in Table 2, B2 has the highest maximum force and is also the best configuration considering the lightweight numbers for elastic and plastic deformation. As the mass differences were 205% between B2 and B1 and 125% between B2 and B3, respectively, we conclude that the shape of B2 is the most promising for being used infill pattern for bending load. Also, the testing showed that the cell's shape has a significant impact on the part behavior. In general, the lightweight numbers are within 10% in the elastic regime, while they differ by nearly 50% in the plastic regime.

Table 2. Results for the maximum force, lightweight numbers and effective density-related-stiffness. The mean values of all specimens are presented in the table

Name	F_{max} in N	M_1	$M_{\max}$	S_{eff} in $\frac{Nmm^2}{kg}$
B1	472	254	681	405,060
B2	1815	282	1276	343,799
B3	765	275	648	424,019

Interestingly, the effective density-related stiffness of B2 is significantly lower than for B1 and B3. Here B3, which corresponds to the length to height ratio of the tested beam specimens, shows the best performance, which leads to ambiguous results depending on the metrics used or to the methodology as such.

Also, the dimensions we used for the infill patterns are unusually large. More convenient sizes would be in the region of 3 to 6 mm. Nevertheless, this step was necessary for two reasons: The software Eiger does not allow self-designed infill patterns, and predefined STL files were too large to be sliced successfully. To provide a comparability with standard infill structures, the use of an open system, which allows the use of self-designed infill patterns is inevitable. Also, the meshing for the FEA becomes enormously costly for these small structures. A solution to this problem could be using representative volume elements and the homogenization approach to use effective material parameters in the FEA.

4 Summary and Outlook

This article presented a basic workflow, which led to optimized infill patterns for predefined load cases. The workflow is based solely on open-source tools and can potentially be combined with slicing software tools. Thus, the approach has a high potential for creating fully automated load case-orientated infill patterns. The FEA shows furthermore that the Hencky model leads to reasonable results. However, there are still some open challenges on choosing an appropriate cell shape and how to select an optimal starting point for the pattern.

For future development, rules and strategies for selecting suitable cell shapes and the arrangement of the unit cells are essential to exploit the full potential of the approach.

Acknowledgements. This study was supported via the European Regional Fund (ERDF). Responsibility for the content of this publication lies with the authors.

References

1. Gibson, I., Rosen, D., Stucker, B., Khorasani, M.: Additive Manufacturing Technologies. Springer International Publishing, Cham (2021).
2. Lachmayer, R., Rettschlag, K., Kaierle, S., Eds.: Konstruktion für die Additive Fertigung 2020. Springer, Berlin, Heidelberg (2021).
3. Bender, B., Gericke, K., Eds.: Pahl/Beitz Konstruktionslehre: Methoden und Anwendungen erfolgreicher Produktentwicklung. Springer, Berlin, Heidelberg (2021).
4. Rosnitschek, T., Hentschel, R., Siegel, T., Kleinschrodt, C., Zimmermann, M., Alber-Laukant, B., Rieg, F.: Optimized One Click Development for Topology-Optimized Structures. Applied Sciences 11(5), 2400 (2021).
5. Mayer, J., Wartzack, S.: A concept towards automated reconstruction of topology optimized structures using medial axis skeletons. In Proceedings of the Munich Symposium on Lightweight Design 2020, pp. 28–37. Springer, Berlin (2020).
6. Rosnitschek, T., Seefeldt, A., Alber-Laukant, B., Neumeyer, T., Altstädt, V., Tremmel, S.: Correlations of Geometry and Infill Degree of Extrusion Additively Manufactured 316L Stainless Steel Components. Materials 14(18), 5173 (2021).
7. Dapogny, C., Estevez, R., Faure, A., Michailidis, G.: Shape and topology optimization considering anisotropic features induced by additive manufacturing processes. Computational Methods in Applied Mechanical Engineering 334, 626 – 665 (2019).
8. Gopsill, J., Shindler, J, Hicks, B.: Using finite element analysis to influence the infill design of fused deposition modelled parts. Progress In Additive Manufacturing 3, 145-163 (2018).
9. Aage, N.: Topology optimization codes written in Python, https://www.topopt.mek.dtu.dk/apps-and-software/topology-optimization-codes-written-in-python, last accessed 2021/9/30.
10. nTopology, https://ntopology.com, last accessed 2021/10/6.
11. Markforged, https://markforged.com, last accessed 2021/10/5.
12. Markforged Onyx data sheet, https://markforged.com/materials/plastics/onyx, last accessed 2021/10/20.
13. Campbell, F. C.: Lightweight materials: understanding the basics Nr. 1. Materials Park, Ohio, USA, ASM International (2012).
14. Klein, B.: Leichtbau-Konstruktion: Berechnungsgrundlagen und Gestaltung. Wiesbaden, Vieweg + Teubner (2009).
15. Xiao, H., Chen, L.S.: Hencky's elasticity model and linear stress-strain relations in isotropic finite hyperelasticity. Acta Mechanica 157, 51 – 60 (2002).
16. Grellmann, W., Seidler, S.: Kunststoffprüfung. München, Hanser (2015).
17. Z88, https://www.z88.de, last accessed 2021/10/19.
18. Wang, C., Zhao, Z., Zhou, M., Sigmund, O., Zhang, X.S.: A comprehensive review of educational articles on structural and multidisciplinary optimization. Struct Multidisc Optim (2021).
19. Harzheim, L.: Strukturoptimierung Grundlagen und Anwendungen (3rd ed.). Verlag Europa-Lehrmittel Nourney, Vollmer GmbH & Co (2019). ISBN 978-3-8085-5882-9.

Topology Optimization and Production of a UAV Engine Mount Using Various Additive Manufacturing Processes

Felix Mesarosch(✉), Tristan Schlotthauer, Marlies Springmann, Johannes Schneider, and Peter Middendorf

University of Stuttgart, Institute of Aircraft Design, Stuttgart, Germany
mesarosch@ifb.uni-stuttgart.de, schlotthauer@ifb.uni-stuttgart.de, springmann@ifb.uni-stuttgart.de, schneider@ifb.uni-stuttgart.de, peter.middendorf@ifb.uni-stuttgart.de

Abstract. Additive manufacturing processes in combination with topology optimization offer great potentials for lightweight design. In this work, the engine mount of a Unmanned Aerial Vehicle (UAV) is topology optimized to minimize weight for a given stiffness. The solutions are carried out taking into account the different material properties of Acrylnitril-Butadien-Styrol-Copolymere (ABS), Polyamide 12 (PA 12), acrylic photopolymer and glass fiber reinforced acrylic photopolymer used for manufacturing. The adapted geometries are then fabricated using Fused Filament Fabrication (FFF), Selective Laser Sintering (SLS) and Stereolithography (SL) to find the most suitable process to produce the weight-optimal solution. In order to validate the simulation results, all parts are tested equally with the same load cases on a customized test rig. The evaluation of the measurements shows that, with one exception, the measured displacements are smaller than the simulation values. By applying a quality criterion, the components made of PA 12 using the FFF process prove to be the weight-to-displacement optimal solution. Compared to the initial design, the topology optimized version saves 40% of mass. In addition, the new design, together with the use of additive manufacturing processes, facilitates the production and assembly of the engine mount.

Keywords: Topology optimization · Additive manufacturing · UAV

1 Introduction

Unmanned Aerial Vehicles (UAVs) are becoming increasingly popular due to their wide range of applications in various fields. These range from small remote-controlled multicopters for recreational purposes to research platforms and autonomously operating military devices. By not requiring a pilot on board, UAVs can often be built small and lightweight or enter areas that would be

J. Rieser et al. (Eds.): *Proceedings of the Munich Symposium on Lightweight Design 2021*, Proceedings, pp. 124–135, 2023. https://doi.org/10.1007/978-3-662-65216-9_12

dangerous for humans. In order to increase flight duration, range or payload ratio, the structures and components of the aircraft should be designed to be as light as possible [1]. Lightweight structures can be created through structural optimization. For this, optimization algorithms accumulate material only at the points where loading occurs [2]. However, the structures created in this way are often very complex and organically shaped and can not be produced with conventional manufacturing technologies. This problem can be solved by additive manufacturing processes, as they can be used to produce geometries of almost any complexity [1].

1.1 Goal

The objective of this work is to improve the engine mount of the experimental UAV called Aircraft for Research and Applied Science (ARAS), which is an aircraft from the Institute of Aircraft Design, University of Stuttgart. The existing engine mount is to be replaced by a topology-optimized structure, which aims to save weight. Various additive manufacturing methods are used for the production. The components are tested on a test rig and the results are compared with the simulation to find out how well the simulation matches reality for the respective processes. Furthermore an assessment shall be given, which process is best suitable for this project.

2 Conceptual Design of the Engine Mount

2.1 Requirements

One of the elementary functions of the engine mount is to hold the engine in its defined place and to introduce all loads occurring during operation into the structure. The engine is a 720 g weighing electric motor TORCMAN NT430-40-Z with 2400 W of power. It is aligned in the direction of the aircraft's longitudinal axis and is located in the rear, which results in a pusher configuration. On the one hand, the engine connection should be as lightweight as possible to increase the range of the UAV. On the other hand, it should have a high stiffness so that no large displacements occur during operation. Further requirements are mountability and inspectability during maintenance work.

2.2 Previous Situation

In the original design, the engine mount consisted of carbon fiber reinforced plastic (CFRP) plates, which were glued to the fuselage shell made of glass-fiber reinforced plastic (GFRP) for reinforcement. The engine was then bolted directly onto the CFRP plates with the fuselage shell. The weight of this engine mount, including adhesive, is estimated at approx. 55 g. This situation is shown in Fig. 1.

2.3 Load Assumptions and Materials

A total of 4 different load cases occur in operation. These are the maximum static thrust, the torque of the motor, the weight force of the motor in the event of a hard landing impact, as well as the resulting bending moment around the connection points of the motor. The loads are summarized in Table 1.

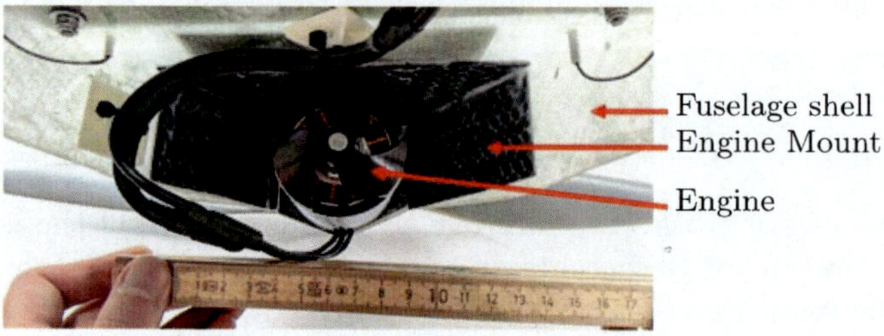

Fig. 1. Previous situation of the Engine Mount

Table 1. Assumed load cases

Thrust	$\vec{F}_{Th} = 100\,\text{N}$
Torque	$\vec{M}_D = 9,88\,\text{Nm}$
Touchdown load	$\vec{F}_{TD} = 148,3\,\text{N}$
Bending moment by touchdown	$\vec{M}_B = 6,578\,\text{Nm}$

It should be noted that the assumed loads represent maximum expected values and are used in the design so that all load cases act simultaneously.

Different materials are used for the various manufacturing processes. Availability, processability on the machines used and costs are decisive factors for their selection. For the FFF process, ABS-M30 and nylon 12 respectively polyamide 12 (PA 12) are used. The SLS process also uses PA 12 with a 50:50 mixing ratio of used to new powder. This allows a comparison of how the different processing methods affect the same materials. For the SL process, acrylic photopolymer is used on the one hand, and acrylic photopolymer reinforced with 20 percent by weight of glass fibers on the other. The material properties used are given in Table 2.

Table 2. Material properties

Material	Young's modulus [MPa]	Poisson's ratio [-]	Density [Kg/m^3]	Yield strength [MPa]
ABS-M30 (FFF)	2180	0.35	1040	26
PA 12 (FFF)	1138	0.41	1000	28
PA 12 (SLS)	1700	0.40	930	48
Acrylic photopolymer (SL)	2700	0.3	1216	74
Glass fiber reinforced acrylic photopolymer (SL)	4287	0.25	1465	79

2.4 Types of Connections

Two different concepts are being investigated for attaching the engine mount to the fuselage shell. One is a glued connection and the other is a screw connection. Gluing is the lightweight option, because no fasteners are required. However, once it is carried out, it can not be removed without destroying the structure. In this regard, the screw connection has a decisive advantage, as it makes it easier to replace the components.

3 Topology Optimization of the Engine Mount

3.1 Design Space

The maximum available design space results from the maximum available space within the fuselage shell. It is delimited in the longitudinal direction because of a circuit board for flight controlling. There is space left out in the design space for the motor and motor ventilation. Excluded from the design space are the three connection flanges to the engine, as these should be present in their entirety. The resulting design space is shown in Fig. 2 in burgundy, the excluded parts are grey.

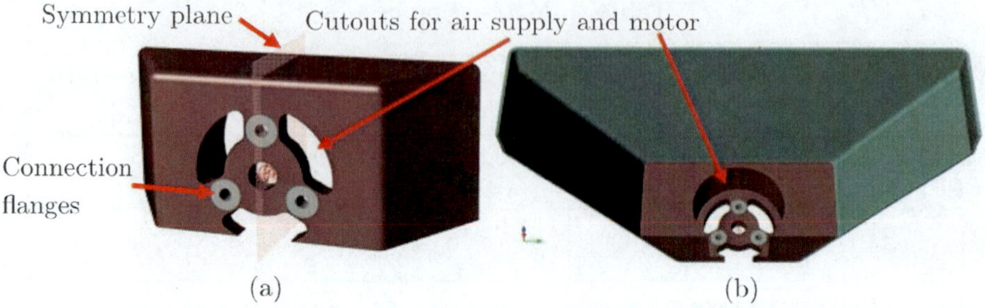

Fig. 2. Design space: without (**a**) and with (**b**) fuselage shell

3.2 Optimization

For the topology optimization the software Inspire from Altair is used. It should be noted that in the approach chosen here, only the different material properties for the respective processes are taken into account. Process-specific peculiarities, such as anisotropy effects due to the layered manufacturing, cannot be included in the optimization. They are taken into account by appropriately reduced material properties, however.

A symmetry plane between the left and right sides of the design space is defined as a boundary condition for manufacturing. It is marked in Fig. 2a as a translucent red plane. It ensures, that the optimization algorithm generates both sides of the mount symmetrically. The load application point is placed in the common center of the three mounting points of the engine and is connected to them via rigid connectors. This distributes the load equally over all connection flanges. As bearing, the entire cross sectional area of the fuselage shell is defined as a fixed bearing. The fuselage shell is just a part of the whole fuselage structure. In order to achieve a more realistic result, it is taken into account in the topology optimization but as structure which is excluded from the optimization. For the connection between the design space and the fuselage shell, the entire contact surfaces are defined as a fixed adhesive joint.

For topology optimization, the objective is to maximize the stiffness for a given mass target. For this, the mass target for the optimization runs is increased from 3% to 12% of the total design area volume in 1% steps. The optimizations are carried out for every material and manufacturing process. The resulting design proposals are then analyzed structurally, in particular with regard to the displacement. In order to determine the optimal solution from the multitude of resulting structures, an evaluation parameter is required. For this, the reciprocal value of the product of the maximum displacement and the weight of the loaded structure is used as a quality criterion [3]:

$$\text{Optimization quality} = \frac{1}{\text{max. displacement} \times \text{weight}} \tag{1}$$

The optimal design proposal according to the quality criterion is smoothed and adjusted constructively. Depending on the type of connection, either the defined adhesive surfaces, or the flanges for the screw connection are added to the structure. Attention must be paid to the mountability of the components, which is why the upper part deviates from the design proposal. Figure 3 shows an example of the design proposal (3a) for the FFF process with ABS-M30 and the derived smoothed structure (3b) for the screw connection.

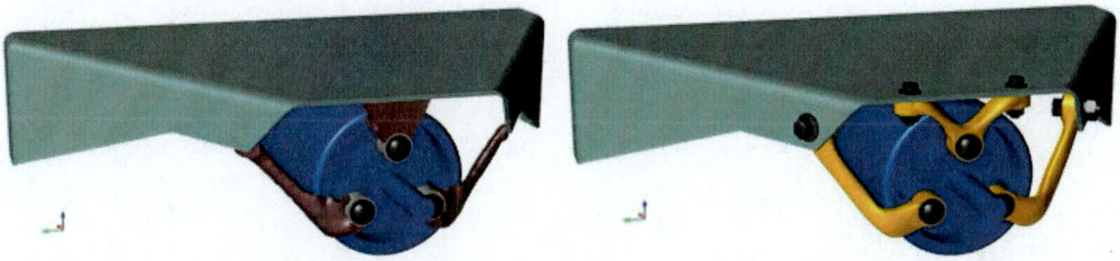

(a) (b)

Fig. 3. Geometry for the FFF process with ABS-M30: (**a**) design proposal; (**b**) post-processed and smoothed structure

4 Manufacturing of the Engine Mount

After topology optimization of the engine mounts, they are manufactured using various additive manufacturing processes. The Stratasys Fortus 450mc machine is used for the FFF process. The layer height is set to 0.254 mm and the infill to 100%. The production of the components with the SLS process takes place on the EOSINT P 390 of the company EOS GmbH with a layer height of 0.12 mm. The parts produced using the SL process are manufactured on the Rapidshape S30L printer with a layer height of 0.1 mm. For the production of the fiber-reinforced components, glass fibers are added to the photopolymer with a weight percentage of 20%. The printer is modified for the manufacturing process by installing a wiper which automatically moves back and forth once after each manufactured layer to prevent sedimentation of the glass fibers. An overview over the processes and materials used is given in Fig. 4.

5 Testing

In order to validate the simulation results, all parts are tested equally with the same load cases on a customized test rig. For this purpose, the part of the fuselage shell used in the simulation is replicated. The loads on the components are applied with turnbuckles, weights, pulleys and monitored with load cells. The deformation of the loaded structures is then measured using the Aramis 4M 3D measuring system from GOM GmbH.

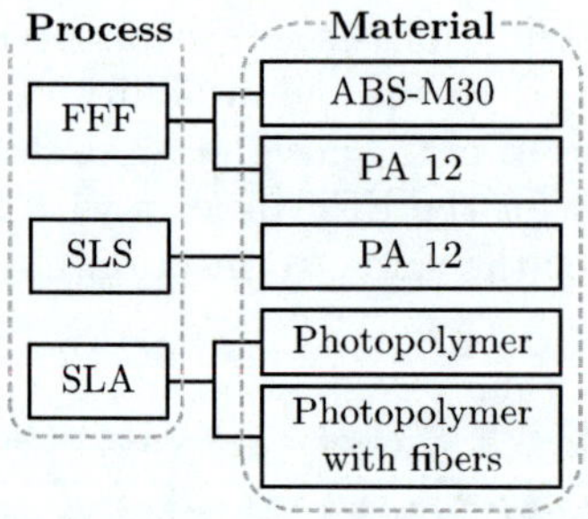

Fig. 4. Overview of the manufacturing processes and materials used

6 Results

6.1 Optimization Quality

For the evaluation of the topology optimization, the quality criterion described in Eq. 1 is applied. The result for the fuselage shell in its initial configuration is shown in Fig. 5. The blue curve represents the quality criterion for the complete design area. To cover smaller mass fractions, the design space must be reduced by 2/3 in the software. This is represented by the red curve. The optimization quality is applied for the maximum displacement occurred in each solution. It can be seen that the optimization quality for the reference shell does not have a global peak. The curve implies that the optimum lies at the boundaries and thus the mass fraction should be reduced further. As the component mass tends towards 0, it can be concluded that the fuselage shell is already sufficiently dimensioned and stiff enough to carry the loads through the engine. Therefore no additional structures are required for the engine mount for this configuration. As a consequence, the entire fuselage shell is included in the topology optimization. The result of this optimization is shown in Fig. 6a.

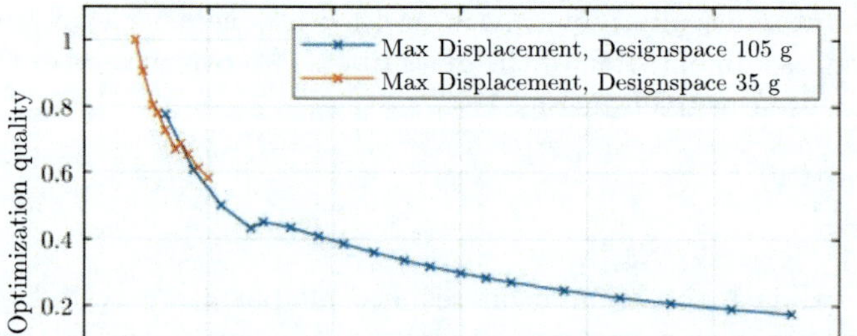

Fig. 5. Optimization quality of the initial fuselage shell - FFF-ABS

As this design would be difficult to manufacture, it is transformed into a simpler one. This can be seen in Fig. 6b. This image also shows the points where the optimization quality is evaluated.

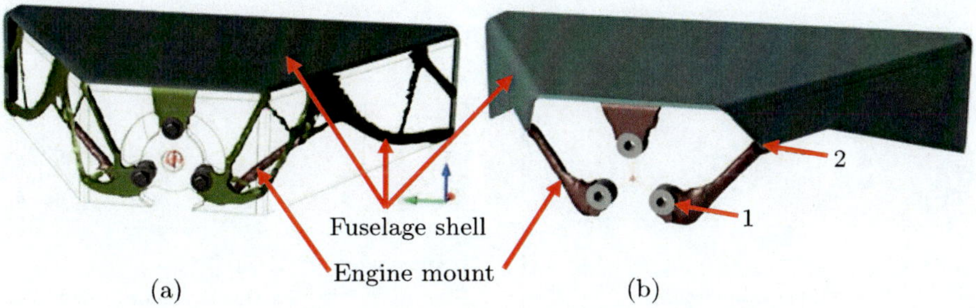

Fig. 6. **a** Design proposal for the topology-optimized fuselage shell; **b** Adapted geometry of the fuselage shell with position of the measuring points

The optimization quality results for the adapted fuselage shell with FFF-ABS are given in Fig. 7.

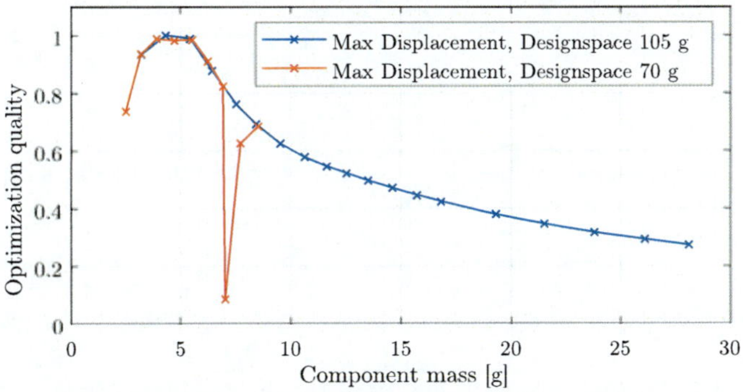

Fig. 7. Optimization quality of the adapted fuselage shell - FFF-ABS

The blue curve is the quality for the complete design space and the red curve represents the space reduced by one third in the initial mass to cover smaller design spaces. The optimization quality in this diagram is applied for the maximum displacement. It can be seen well that in the area of approx. The part resulting from this investigation is manufactured for comparison with all materials and processes and is subsequently referred to as the reference geometry.

Analogous to the reference geometry, the design of a geometry adapted to each material is carried out, as can be seen in Fig. 4. The results of the optimization quality for these optimizations are shown in Fig. 8 for each process and material.

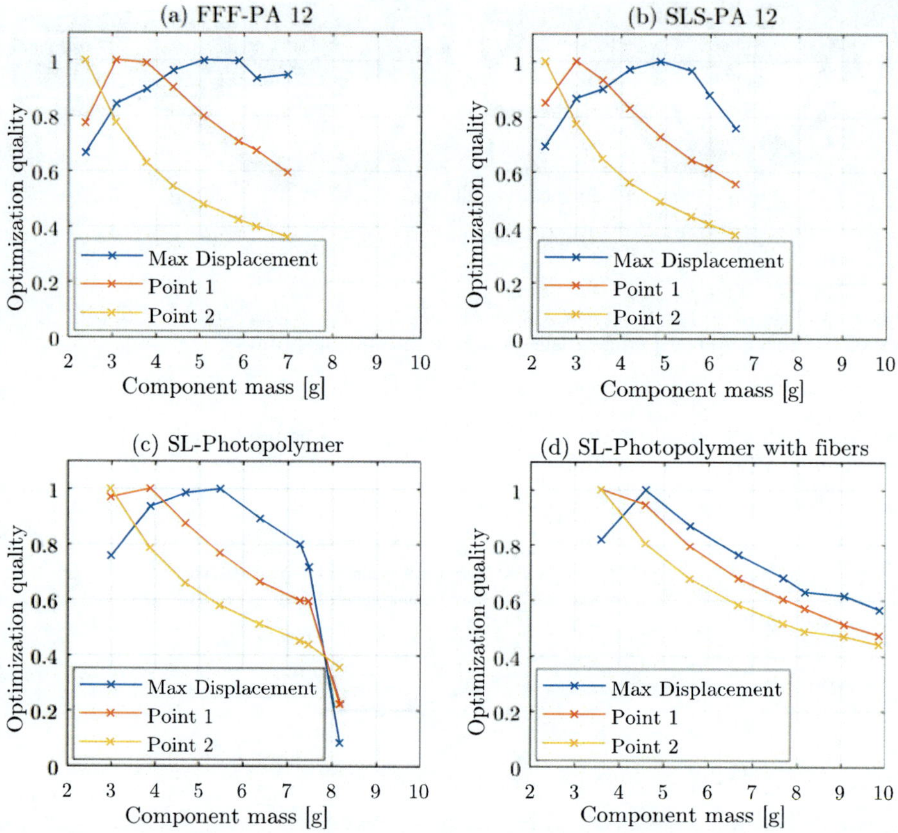

Fig. 8. Optimization quality of the adapted geometries

The points at which the optimization quality is evaluated are drawn in Fig. 6b. The course of the curves for the three points under consideration shows a similar behavior in all four diagrams. The curve of point 2 does not provide an optimum in all cases. This is due to the fact that the point is located on the shell and therefore the displacement decreases only relatively slightly with increasing component mass.

For PA 12, point 1 delivers a peak for the FFF and SLS process each at 3 g, for the maximum displacement it is in the range of 5 g for both methods. Since the geometry of the design proposal for the optimum of the maximum displacement hardly differs from each other for both methods, the same smoothed geometry is used for the further investigations.

For the SL process with photopolymer, the optimum of point 1 is also around 3 g. The peak for the maximum displacement is located at 5.5 g. The geometry of the maximum displacement is used for further investigations.

In the SL process with fiber-reinforced photopolymer, there is no optimum for point 1 because the initial mass of the design space is too large for this and smaller component masses cannot be achieved. The maximum displacement provides a peak at a component mass of approx. 4.5 g. The optimized structure of this mass is smoothed for further use.

6.2 Comparison of the Measurements with the Simulation

The comparison between the measurements under load on the test rig and the simulation is shown in Fig. 9 for the mass (a), the displacement (b) and the normalized quality (c).

It can be seen that the glued variant is most lightweight in both simulation and measurement. This is due to the fact that the flanges for the screws are omitted for these parts. Furthermore, all components except those produced with the SL process are lighter in measurement than predicted by the simulation. This is due to the fact that with the FFF and SLS process, there are small voids in the components, but this does not apply to the SL process.

If the displacements are considered, it is noticeable that there are no measured values for FFF-ABS-glued and SLS-PA 12. This is caused that the adhesive bond broke during the tests and the sintered parts are produced with defects due to technical problems of the machine. Furthermore, it should be noted that the measured values for all parts are smaller than the simulation values and are relatively constant in the range of 1.5 mm. Only for the adapted geometry with SL-photopolymer, the measured value is larger than the simulation, which is probably owed that the simulation underestimates the displacement for this part. The reason for this cannot be clearly identified. One possible cause could be the mesh discretization. In the software used, the user has no influence on this.

Regarding the quality calculated according to Eq. 1, it should be noted that the components with FFF-PA 12 are the best solution according to the measurements. Considering the simulation values, the components made of SL-photopolymer represent the best solution.

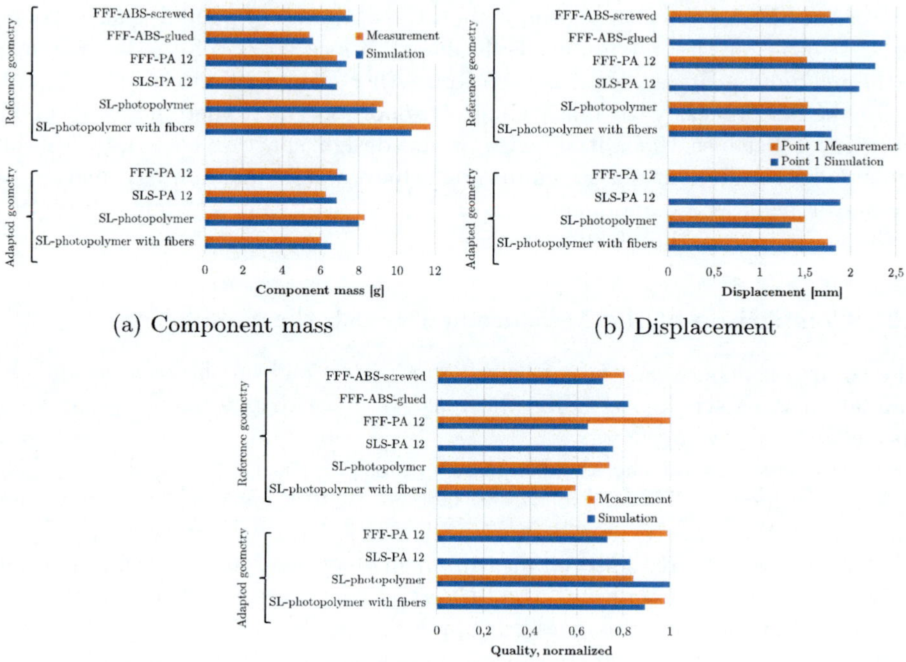

(a) Component mass (b) Displacement

(c) Quality, normalized

Fig. 9. Comparison of the measurements with the simulation

7 Conclusion

Compared to the previous situation, which is shown in Sect. 2.2, Fig. 1, a weight saving of approx. 80 g can be achieved by optimizing the fuselage shell and the engine mount. In relation to the total mass of the examined part of the fuselage shell including the engine mount in the previous design, this corresponds to a reduction of 40%. This weight saving illustrates the potential of additive manufacturing processes in combination with topology optimization. While the simulation cannot accurately represent the processes and materials, the measurements show that the models and assumptions are conservative and provide slightly larger deformations. For these reasons, the process chain consisting of topology optimization, simulation and additive manufacturing for the production of lightweight structures in the UAV area can be recommended. The additional effort is justified by the weight savings, as this brings a great advantage with an increase in payload or range. Furthermore, the manufacturing of the fuselage shell is significantly simplified by the optimization. The new design, together with the use of additive manufacturing processes, facilitates the production and assembly of the engine mount.

References

1. Goh, G., Agarwala, S., Goh, G. L., Dikshit, V, Sing, S. L., Yeong, W. Y.: Additive manufacturing in unmanned aerial vehicles (UAVs): Challenges and potential. In: Aerospace Science and Technology 63 (2017), S.140-151. – ISSN 12709638
2. Schumacher, A: Optimierung mechanischer Strukturen. Springer, Berlin Heidelberg (2013).
3. Sauer, A.: Bionik in der Strukturoptimierung: Praxishandbuch für ressourceneffizienten Leichtbau. 1. Auflage. Würzburg: Vogel Communications Group (2018)

Multiparametric Design Optimisation of 3D Printed Aircraft Door Seals

Bruno Franke Goularte, Vivianne Marie Bruère(✉), Alexander Lion, and Michael Johlitz

Bundeswehr University Munich, Neubiberg, Germany
bruno.franke@unibw.de, vivianne.bruere@unibw.de, alexander.lion@unibw.de, michael.johlitz@unibw.de

Abstract. Additive Manufacturing is a young, promising manufacturing method that is currently coming into focus for its faster and relatively cheaper fabrication of complex, custom-made parts in comparison to conventional manufacturing methods. Prototyping is one of the most favoured fields, as on-demand 3D printing of components can be particularly beneficial at the Research & Development stage. Among the employed materials, one can find elastomer alternatives as traditional liquid silicone rubber and photopolymers. This work deals with mechanical investigation of four silicone technologies, with emphasis on the material characterization. Hyperelastic laws were fitted to tensile tests for Finite Element Method simulations of aircraft door seals. A proposal for multiparametric design optimisation of the seal geometry through genetic algorithms (NSGA-II) is presented based on analysis results. The potential application of each 3D printing technology in seals prototyping is highlighted and evaluated according to the collapse criteria under cabin pressure. Results indicate the variability of the optimal parametric designs according to both the silicone behaviour and the material model stability limitations. The impact of the material model is also underlined as a way to enable seal design improvements.

Keywords: Prototyping · Elastomer · Material characterization · NSGA

1 Introduction

The use of Additive Manufacturing (AM) comes to evidence more and more as a reflection of the advantageous fabrication of parts on site, on demand. Although the current trend is heading towards rapid manufacturing, it is still an interesting ally for prototyping in the industry. Particularly for the aerospace sector, the adoption of AM leads to benefit in the development stage by combining it with

J. Rieser et al. (Eds.): *Proceedings of the Munich Symposium on Lightweight Design 2021*, Proceedings, pp. 136–153, 2023. https://doi.org/10.1007/978-3-662-65216-9_13

numerical simulations for the behaviour prediction of functional parts. While the manufacturing of new parts could take 6–12 months to be accomplished, including the ordering of the geometry, mould fabrication, production itself and final delivery, the 3D printing of a small number of complex parts can be performed from the 3D model data without the use of individual tooling, with reduced waste by-products. Moreover, depending on the geometry size and technology employed, AM requires only a few hours to days without associated delivery time and costs [4,10,19,21]. The faster fabrication of such components may assist on the evaluation of their design and mechanical performance in operation, identification of sources of errors in the early stages and, consequently, provide further improvements on conceptual design optimisation and reduction on laborious and time-consuming numerical analyses of aircraft structures.

In aerospace applications, the majority of AM involves metals, mostly powder-based technologies for titanium and nickel alloys, due to the high performance requirements [20,22,25,28]. Nonmetallic parts are employed as well to some extent in rapid prototyping, tooling for the production of composite parts, fixtures and interior components [20,26]. AM is notably beneficial in this sector since there are lower batch sizes than in other industries, like the automotive, also being an interesting alternative for the just-in-time production of parts for repairs while reducing maintenance and inspection times of the aircrafts unable to be operated [25,28].

The main limitations for the full implementation of AM for manufacturing purposes are the rigorous certification requirements, the standards for testing and the safety and maturity of the technologies [20], although several advances are taking place in the last years. The lack of standard material properties and performance database also hinders the designing of components by engineers and designers, besides the need for process control in order to provide consistency and reliability in the manufacturing. Prototyping, on the other hand, does not have such high demands. Therefore, AM can be put to more use, helping in reducing the time and costs related to the product development cycle [26] regarding the form and appearance and, as the technologies and materials evolve, the functionality of the parts.

Associated to the further exploration of AM is the understanding of the printing processes and modelling to assist in the prediction of material properties. A great portion of current studies involving numerical analysis and AM consists of investigations on the printing process for its optimisation and consequent quality improvement of the final part. Common topics involve the use of the Finite Element Method (FEM) to model metal deposition heat transfer [27], as well as in combination with optimisation algorithms [2], for powder bed fusion processes. It is furthermore used for the thermomechanical analysis for residual stress and distortion predictions in direct energy deposition [7]. FEM is also applied in the analysis of material extrusion technologies for predictions of the mechanical behaviour [3], evaluation of damage under compressive loads [12], quantification of the effect of microstructural imperfections [29], and predictions of printing parameters on layer adhesion [5,24].

A more recent topic of study is focused on elastomers fabricated by AM. They can be of great value for the aerospace industry in sealing applications. Although those are not of first priority for the operation of aircrafts, a proper seal design choice can reduce risks of leakage [11]. Moreover, conventional prototyping is time-consuming and does not allow the investigation of more design options as a consequence of the strict product development and certification deadlines, in which AM can be decisive. Most of the current AM technologies are not suitable for the use of traditional rubber, and vulcanisation cannot be easily transferred to printing processes. Nevertheless, a few of them are able to print with conventional Liquid Silicone Rubber (LSR). For other technologies, new materials with rubber-like properties are developed to fulfil the process requirements. Alternatives include photopolymers for vat photopolymerization, material jetting and binder jetting printers (cured by Ultraviolet - UV - light), as well as thermoplastic elastomers for powder bed fusion and material extrusion machines, which require the use of thermoplastics.

Numerical analyses of 3D printed elastomers is a still modest research subject that has plenty of potential. In any case, in order to make use of Finite Element Analysis (FEA) tools on the evaluation and optimisation of not only the printing process itself but also the final printed parts, material characterization is an important step to be performed [15,16]. The implementation of constitutive models into commercial FE software facilitates the numerical approach and provides more realistic results, assuming that constraints should be imposed due to material stability [23].

With this in mind, this work deals with the parameter identification for 3D printed silicone materials from different AM technologies, according to available material models with a polynomial free energy function with second-order terms [18] and respecting material stability criteria [13,14]. Information about the material behaviour of silicone is provided by uniaxial tensile tests, while the optimisation of a parametric door seal design is carried out with genetic algorithms [30]. The aim is to find the best geometry employing current AM technologies of silicone, by minimising the reaction forces of the door closure and the system weight. Similar studies were carried out in the automotive industry [31]. Although topological optimisation may be addressed in works related to contact problems [9], nonlinear FEM simulations are prone to numerically diverge, making the use of nonparametric techniques quite challenging. By considering large deformations of door seals under cabin pressure, the presented analyses attempt to represent a straight cross section prototype in plane strain state, while avoiding structural and material instabilities.

2 Overview on the Investigated AM Materials

2.1 3D Printing Materials and Processes

The four selected materials consisted of two silicones, named here as "Silastic" and "Wacker", and two grades of a photopolymer with silicone in their composition, named "Keyence-H" and "Keyence-L". More detailed information is found in Table 1 for each type of silicone used in this study.

Table 1. Material and printing information for the investigated silicone materials

Material	Silastic	Wacker	Keyence-H	Keyence-L
Technical name	SILASTIC™3D 3335 LSR	ACEO® Silicone GP	AR-G1H	AR-G1L
Shore Hardness	50 A	50 A	65 A	35 A
Company	Dow (USA)	Wacker Chemie AG (Germany)	Keyence (Japan)	
AM technology	LAM	DOD	MJ	
3D Printer	innovatiQ LiQ 320	ACEO® Technology	Keyence Agilista-3000	

Silastic is a conventional two-part, thermo-cured LSR printed with Liquid Additive Manufacturing (LAM) from innovatiQ (Germany). The silicone is extruded through a nozzle and a 2000W halogen lamp provides the heat for curing. Flat plates of $60 \times 60 \times 2$ mm were printed, from which dogbone samples were cut for posterior mechanical testing (see Appendix). Wacker is an UV-cured silicone printed in Wacker's trademark Aceo 3D printing technology. Since Aceo is a paid service provider, details on the printing parameters are not available. It is known, however, that it is a process based on Drop-On-Demand (DOD) with water-soluble support material. It uses inorganic synthetic rubber, which undergoes a vacuum post-curing at 200°C for 4 h to improve the mechanical properties and remove the volatiles [1]. Dogbone-shaped samples were acquired from this technology. Finally, Keyence-H and Keyence-L are elastic photopolymers printed in Keyence's Material Jetting (MJ) process without the need of post-curing. The materials are especially developed for the technology itself (closed source); the type of material is chosen in the slicing software and no printing parameters need to be manually set. Both materials possess the same chemistry, while the component content gives the different hardness grades and elastic capabilities. As well as for Wacker, dogbones were directly printed.

The choice of materials was based on the fact that current aircraft sealing components are made of silicone. Moreover, these printing processes avoid the use of a greater range of printing parameters by the printer operator that might influence the properties of the final part, reducing the variability in the results. This becomes convenient, as the focus of this study is not the improvement of the printing process.

2.2 Uniaxial Loading

Tensile tests were performed for 5 specimens per material on a Zwick-Roell© testing machine at room temperature for a force sensor of 500 N and a strain rate of 15 mm/min. Dogbone geometries according to DIN 53504-S3A [6] were pulled up to 100 % strain, or stretch $\lambda = 2$, shown in Fig. 1.

(a) Silastic (b) Wacker

Fig. 1. Uniaxial loading experiment performed on Zwick-Roell machine

Figure 2 presents the average results converted to the first Piola-Kirchoff (PK1) stress, i.e. engineering stress, for each silicone type. It can be noted that the silicones from Silastic and Wacker have very similar behaviour and values, which is expected as Silastic is a typical LSR and Wacker is a "100% pure silicone", according to the manufacturer. Keyence-H and Keyence-L behave in a more distinct way, with the differences in the stress intensities as a reflection of the lower/greater elastic composition. The mechanical properties are a direct consequence of the material chemistry and the AM process nature.

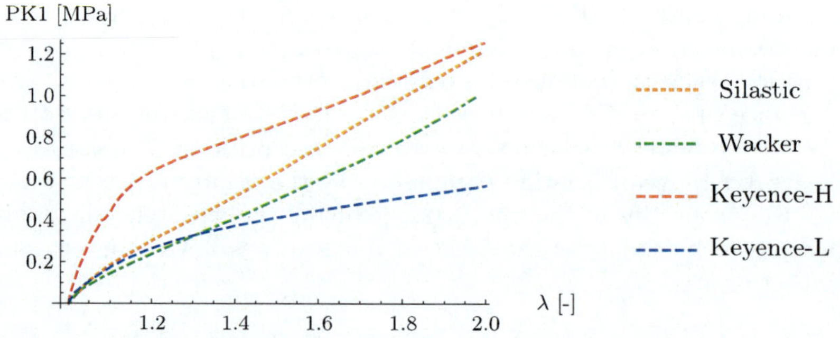

Fig. 2. Stress-strain curves for uniaxial loading of each technology

3 Mechanical Behaviour of Silicone

After performing tensile tests, the material parameters for each specimen were identified according to their mechanical response up to 100% strain, or $\lambda = 2$. This study is based solely on the hyperplastic behaviour of the different 3D-printed technologies to prototype seal designs through 2D simulations assuming plane strain states around the straight sections of the door contours. Thus, it is important to investigate which hyperelastic models are already implemented in commercial FEM software for easy design optimisation, without requiring the development of user defined models.

3.1 Hyperelastic Incompressible Models

Assuming homogeneity and isotropy in the specimens, the elastic behaviour is approximated through the choice of free energy functions that are consistent with the stress-strain response of the tensile tests. By adding the constraint of incompressibility for the silicone, the deformation gradient $\mathbf{F}$ is written as a function of the uniaxial stretch λ, imposing that $\det[\mathbf{F}] = 1$:

$$\mathbf{F} = \begin{pmatrix} \lambda & 0 & 0 \\ 0 & \frac{1}{\sqrt{\lambda}} & 0 \\ 0 & 0 & \frac{1}{\sqrt{\lambda}} \end{pmatrix} \tag{1}$$

The models are then formulated based on the first and second invariants of the left Cauchy strain tensor $\mathbf{B} = \mathbf{F} \cdot \mathbf{F}^T$:

$$I_1 = \text{tr}[\mathbf{B}] \tag{2}$$

$$I_2 = \frac{1}{2} \left(\text{tr}[\mathbf{B}]^2 - \text{tr}[\mathbf{B} \cdot \mathbf{B}] \right) \tag{3}$$

The generalized Mooney-Rivlin free energy function is initially considered since it is already implemented in commercial FEM software [17,18], assuming the following form:

$$\psi(I_1, I_2) = \sum_{i+j>1}^{n} C_{ij} (I_1 - 3)^i (I_2 - 3)^j \tag{4}$$

where C_{ij} are the coefficients of the n-th order polynomial free energy function depending on the strain invariants. The Cauchy stress tensor $\mathbf{T}$ can be written by adding the pressure term p that satisfies incompressibility.

$$\mathbf{T} = 2 \left[\left(\frac{\partial \psi(I_1, I_2)}{\partial I_1} \right) \mathbf{B} - \left(\frac{\partial \psi(I_1, I_2)}{\partial I_2} \right) \mathbf{B}^{-1} \right] - p\mathbf{1} \tag{5}$$

In order to compare the tensile tests in terms of engineering stresses, the 1st Piola-Kirchoff tensor $\mathbf{P}$ is calculated:

$$\mathbf{P} = \mathbf{T} \cdot \mathbf{F}^{-T} \tag{6}$$

3.2 Parameter Identification

The identification of the model parameters was performed by means of the numerical optimisation algorithm *NMinimize* in Wolfram Mathematica 12, by minimizing the stress difference between model and experiment:

$$MSE = \frac{1}{N}\sum_{k=1}^{N}\left(\frac{P_k^{model}(\lambda_k) - P_k^{experiment}(\lambda_k)}{P_k^{model}(\lambda_k)}\right)^2 \tag{7}$$

This function expresses the sum of the mean squared errors (MSE) between the experimentally measured stress values from N-data points and the model output at each stretch value depending on the parameters of the free energy function. Among the internal conditions of the material model, Drucker's stability criterion [8] is used to ensure that for each stretch increment, the free energy function is always increasing. It is necessary that the free energy function also covers other loading states, remaining polyconvex at least for a limited range of stretches. To consider different stress-strain modes other than uniaxial, this criterion is verified for the biaxial tension and compression with independent stretches λ_1 and λ_2. To this end, the deformation gradient is rewritten as:

$$\mathbf{F}_{biaxial} = \begin{pmatrix} \lambda_1 & 0 & 0 \\ 0 & \lambda_2 & 0 \\ 0 & 0 & \frac{1}{\lambda_1\lambda_2} \end{pmatrix} \tag{8}$$

For the identified material model, Drucker's stability criterion should be satisfied at the change $d\mathbf{T}$ in the stress driven by an arbitrary logarithmic strain increment $d\varepsilon$ by the following inequality:

$$d\mathbf{T} : d\varepsilon > 0 \tag{9}$$

Using the material tangent $\mathbf{D}$ with the relation $d\mathbf{T} = \mathbf{D} : d\varepsilon$ in the inequality:

$$d\varepsilon : \mathbf{D} : d\varepsilon > 0 \tag{10}$$

It is required that $\mathbf{D}$ must be positive-definite for any values of strain. Considering for the plane stress condition $T_{33} = 0$, the material tangent can be written in matrix form with respect to the main stretches using the relation between the logarithmic strains $d\varepsilon_{11} = d\lambda_1/\lambda_1$ and $d\varepsilon_{22} = d\lambda_2/\lambda_2$:

$$\mathbf{D} = \begin{pmatrix} D_{11} & D_{12} \\ D_{21} & D_{22} \end{pmatrix} \begin{pmatrix} d\lambda_1/\lambda_1 \\ d\lambda_2/\lambda_2 \end{pmatrix} \tag{11}$$

Thus, the material stability check is performed for values of λ_1 and λ_2 under the limits of tension and compression which are relevant for aircraft seals. From the definition of a positive-definite matrix, we find:

$$\mathrm{Tr}[\mathbf{D}] = D_{11} + D_{22} > 0 \tag{12}$$

$$\det[\mathbf{D}] = D_{11}D_{22} - D_{12}D_{21} > 0 \tag{13}$$

Both inequalities are used to evaluate the quality of the identified parameters for loadings that are not represented by the uniaxial test. Table 2 presents the fitted parameters for a polynomial energy function of second order according to Eq. 4. The results show that the Silastic and Wacker silicones presented good correlation while presenting also material stability. All fitted parameters came positive. On the other hand, both types of photopolymers from Keyence could not satisfy the stability criteria for a good fit without having some negative fitted parameters, which are discussed in Subsect. 3.3.

Table 2. Identified parameters with 2nd order law according to each technology

Model Silicone	MSE [%]	C_{01} [MPa]	C_{10} [MPa]	C_{11} [MPa]	C_{02} [MPa]	C_{20} [MPa]	Stability check
Silastic	2.10	0.345	0.000	0.079	0.004	0.000	Valid
Wacker	4.63	0.279	0.000	0.019	0.000	0.028	Valid
Keyence-H	8.87	0.831	0.000	-0.090	-0.084	0.085	Limited
Keyence-L	6.85	0.319	0.000	-0.030	-0.040	0.030	Limited

Figure 3 presents the first Piola-Kirchoff (PK1) stress data obtained in the tensile tests and the fitted models as function of the stretch λ. The dotted curves represent the experimental data, while the solid ones are the models.

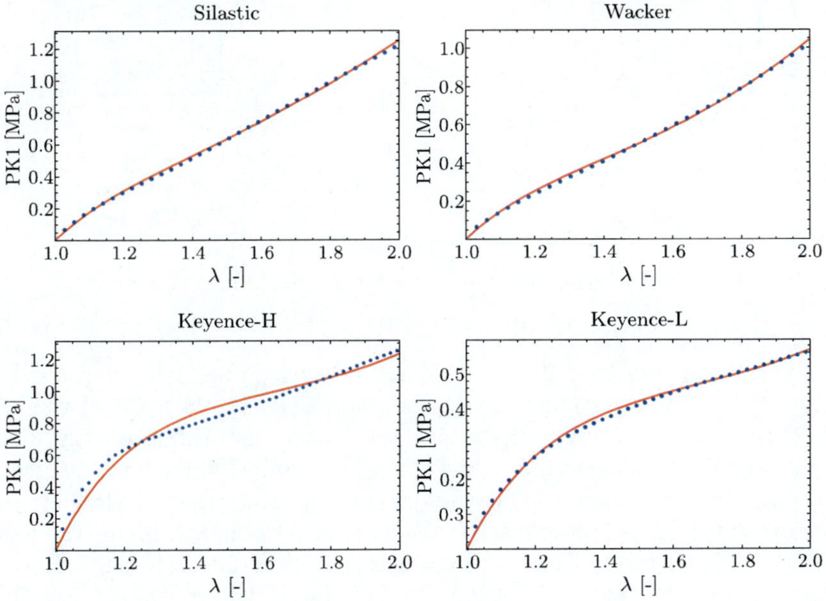

Fig. 3. Fitted curves of each technology (dotted - experiment / line - model)

3.3 Model Stability Limitations

Due to the change in the stiffness of the Keyence materials from 20% deformation onwards, a different free energy function should be considered to avoid negative parameters and to render the mechanical behaviour more consistent. As this study is restricted to material laws that are already implemented in the ABAQUS software [17], the Drucker's stability condition in the biaxial case is used to bound the identified parameters within expected limits for door seals. For this purpose, stretch limits for λ_1 and λ_2 are limited between 0.65 and 1.35, which takes into account the compression and tension limits of the seal in operation, respectively. This boundary is considered for the seal optimisation on Sect. 4. By taking into account the model limitations, the material stability criterion is illustrated in the regions of interest. Figure 4 shows how the strain energy is restricted for some stretch combinations. Despite not respecting the polyconvexity [13,23] for all stretch levels, the material model for the Keyence silicones must be limited for seal loading conditions, remaining stable inside the green mashed region.

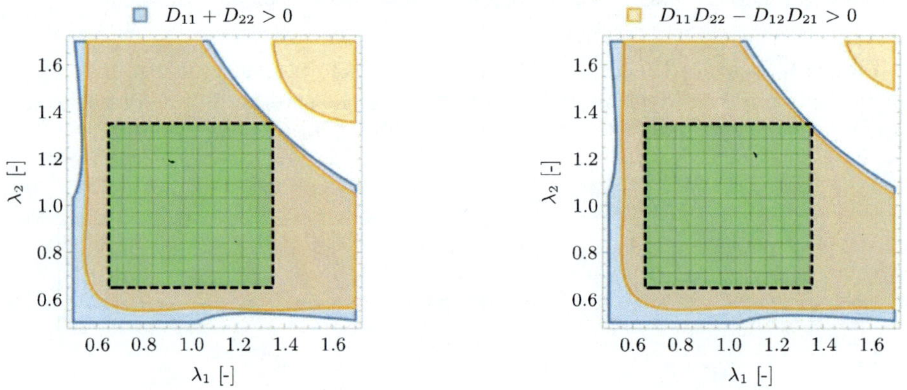

(a) Keyence-H

(b) Keyence-L

Fig. 4. Free energy function domain valid for Drucker's stability

More experiments should be considered to suit different loading conditions and to fit the model parameters, e.g. simple shear. In the case of the stress-strain curves of the Keyence materials, the choice of the free energy function may also impair the correct modelling of the mechanical behaviour of silicone. It is observed that the stress response under uniaxial loading suits the polynomial energy order for Silastic and Wacker elastomers. Although it is not the focus of this work, it is desirable that the hyperelastic model for Keyence silicones are able to capture the material softening under large deformations. Lower order free energy functions without employing negative parameters that multiply the strain invariants may avoid the use of stretch restrictions considered in the seal optimisation procedure.

4 Seal Designing Optimisation

After identifying the parameters of the hyperelastic models for each type of silicone, seal simulations were carried out by employing the different 3D printing technologies. Based on a previous study with seal applications for aircraft flight cruising conditions [11], a FEM boundary value problem is proposed considering the cabin pressure and the contact conditions for a simple 2D plane strain model. The seal geometry is parametrised for a prototype seal-striker interface with the dimensions presented in Fig. 5:

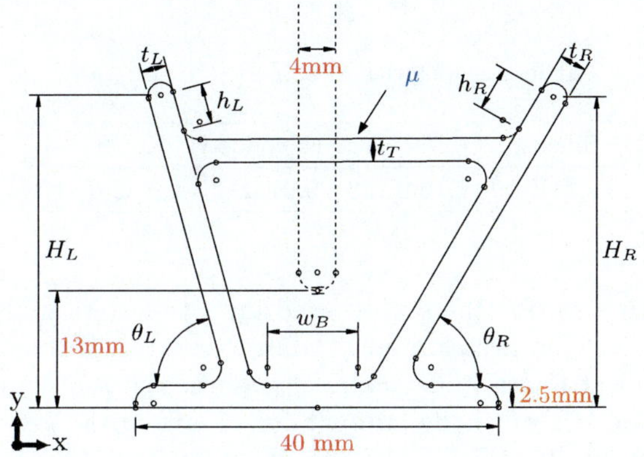

Fig. 5. Seal prototype (solid) and rigid striker (dashed) cross sections

Table 3 shows the range in which these cross section dimensions are used in the context of door seals. Some of the interface values are prescribed in order to establish a baseline for the design optimisation: a door striker closed gap of 13.0mm, the seal bottom dimensions fixed as showed and fillet radii set to 2.0mm for the geometry. Since low friction layers can be added to improve the surfaces of the 3D printed seals, the coefficient of friction is also considered. Despite being made of fabric materials, the influence of such thin layers on the global mechanical behaviour of the seal is disregarded in this study.

Table 3. Design parameters range of the door seal optimisation

Design parameters	Description	Minimum	Maximum
t_L, t_R, t_T	Thickness (seal wall)	1.0mm	5.0mm
H_L, H_R	Height (seal wall)	35.0mm	43.0mm
h_L, h_R	Height (lip feature)	1.0mm	5.0mm
θ_L, θ_R	Angle (seal wall)	60°	90°
w_B	Width (seal bottom)	5.0mm	15.0mm
μ	Friction coefficient	0.1	0.4

4.1 Optimisation Tasks

To initiate the design optimisation process, objective functions to be minimised are needed. The lowest possible reaction force (RF) for the cross-section compression is sought so that the aircraft door does not require more robust parts for the latching mechanism. In addition, the aim is to reduce the weight of the seal itself in order to minimize the mass of the system as a whole. In this way, each type of material is evaluated based on its mechanical characteristics in the search of an optimum geometry that delivers the best seal performance. Table 4 presents the measured densities of each material employed for the optimisation.

Table 4. Densities of silicones in ton/mm^3

Silastic	Wacker	Keyence-H	Keyence-L
1.121E-09	1.164E-09	9.978E-10	9.824E-10

A stability criterion for the seal is used on a two-step analysis for the optimisation tasks (Seal Compression and Cabin Pressure). If the cross section does not offer sufficient stiffness under cabin pressure, the seal can collapse. Structural instability occurs once the contact forces are no longer able to hold the cabin pressure ($\Delta P = 0.0465\,\text{MPa}$) on the contact boundary, due to insufficient tangential forces driven by friction. This failure criterion is defined from the model outputs based on relative displacements between striker $U_x^{striker}$ and seal lip U_x^{lip} in the x-axis direction. Figure 6 illustrates a frame previous to the seal collapse in the second step of the simulation. The upper right lip slips under the rigid interface to the atmosphere side, which does not enable the seal to recover its original deformed configuration in the first step. Taking into account the two-step analysis results, an optimisation task was defined to search for the best geometric parameters of the seal while avoiding collapse.

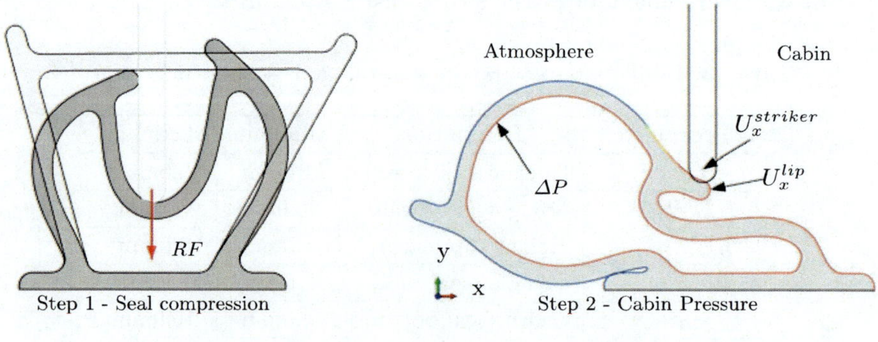

Fig. 6. Seal about to collapse in a two-step FEM analysis

4.2 Multi-objective Search with NSGA-II

Given the problem of minimizing the door closure reaction forces (RF) and the mass per metre of seal length (M), a multi-objective algorithm was assigned to find "the best" design solutions. Using the software iSight, the Non-dominated Sorting Genetic Algorithm (NSGA-II) is applied to find the best candidates with a number of 20 generations containing populations of 12 seal simulations, in total 240 runs. Despite of heaving different dimensions and magnitudes, the algorithm tries to improve both outputs independently, without dominance by one of the objectives. By means of genetic mutations and crossover between the design variables (x_i), the "best" runs are selected according to the improvements from previous runs, establishing a Pareto Front. The following scheme presents how the multi-objective problem is defined:

$$\min_{x_i} \; f(x_i) = \omega(RF(x_i) + M(x_i)) \tag{14a}$$

$$\text{s.t Seal collapse} \rightarrow \; U_x^{lip} > U_x^{striker}, \tag{14b}$$

$$\text{Material stability} \rightarrow \; 0.65 < \{\lambda_1, \lambda_2\} < 1.35, \tag{14c}$$

$$x_i \in \text{Table 3} \tag{14d}$$

The parameter ω is a weight that penalises the multi-objectives if the constraints are violated, the so-called unfeasible runs. While condition 14b refers to the seal collapse fail criteria, the condition 14c is only assigned to Keyence materials to keep material stability. The penalty function $f(x_i)$ represents the sum of contribution of the objectives to be minimized. Figure 7 illustrates an example of optimisation considering the Keyence-H silicone. The boundary of the trade-offs between seal mass and the door reaction force is seen as a Pareto Front, where it is not possible to change the design variables x_i without affecting one of the objectives, at least.

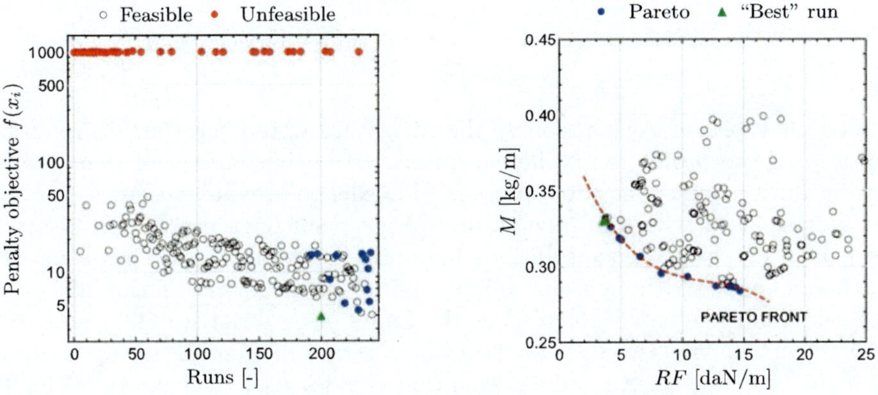

Fig. 7. Multi-objective design optimisation example with Keyence-H silicone

4.3 Optimisation Results

Once the optimisation tasks were completed for each type of silicon, the "best" design solutions are compared according to their multi-objectives of reaction force and mass per seal length. Figure 8 illustrates the results belonging to the deformed configuration under cabin pressure for the four seal designs with maximum absolute logarithmic strain (ε) contour plots.

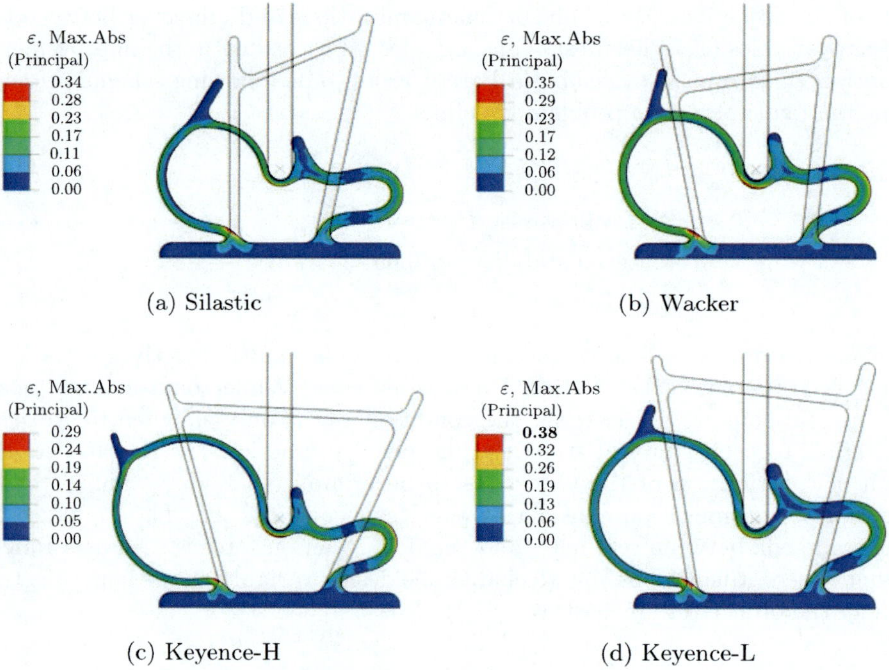

Fig. 8. Different seal designs found by the optimisation algorithm (NSGA-II)

For all types of AM silicones, the algorithm opted for thin walls on the majority of the profile, with the exception of the right side wall that requires more stiffness against the cabin pressure. In order to provide structural stability, the designs used with the Silastic and Wacker silicones present profiles with wider angles. The Keyence solutions maintained a higher aspect ratio between width and height. Although seal collapse did not occur for the all candidates, the Keyence-L silicone could not respect the limits of material stability. Due to its mechanical behaviour which is less stiff than the other materials, the optimisation algorithm was not able to find a design that could satisfy the condition 14c. The following expression converts the logarithmic strains to stretches:

$$\lambda = \exp(\varepsilon) \tag{15}$$

Despite presenting larger lateral displacements due to the seal inflation by the pressure, the solution found using the Keyence-H silicone respected both structural and material stability conditions. Table 5 presents the combined design variables that enabled to keep the seal in a stable configuration while minimizing the objectives for each silicon printed by a specific AM technology.

Table 5. Multi-objective results for each silicone AM technology

Design parameter (x_i)	Silastic	Wacker	Keyence-H	Keyence-L	Unit
Friction coefficient (μ)	0.245	0.335	0.192	0.355	-
Angle left (θ_L)	89.8	79.3	74.6	81.0	degrees
Angle right (θ_R)	78.6	84.3	65.7	76.1	degrees
Height left (H_L)	36.5	35.2	36.9	41.7	mm
Height right (H_L)	39.8	36.3	35.0	39.8	mm
Lip height left (h_L)	3.7	4.2	1.3	1.1	mm
Lip height right (h_R)	1.1	1.9	2.0	3.2	mm
Wall thickness head (t_H)	1.2	1.6	1.3	1.6	mm
Wall thickness left (t_L)	1.6	1.4	1.1	1.6	mm
Wall thickness right (t_R)	2.0	2.3	1.9	2.5	mm
Width bottom (W_B)	8.8	5.3	12.5	14.2	mm
Reaction force (RF)	12.45	10.04	5.12	8.41	daN/m
Seal mass (M)	0.299	0.320	0.267	0.314	kg/m

The results show that the designs based on the Keyence-H silicone can considerably reduce reaction forces and system mass through a suitable combination of mechanical and geometrical characteristics. During the seal compression step, due to a lack of contact between the seal lips and the rigid striker combined with a relatively lower friction coefficient, the best design among the other technologies was able to take advantage of the stiffer mechanical characteristics of a harder silicone. However, other hyperelastic models that are more consistent with the mechanical behaviour of the Keyence silicones could contribute to less constrained optimisation tasks, not influenced by material modelling instabilities. Table 6 shows that the optimisation time was lower for the Silastic and Wacker silicones for each run and the total working time compared to the Keyence materials. The optimisation tasks were limited to run 5 parallel simulations simultaneously, using 2 cores from the CPU.

Table 6. Statistics for elapsed time of the 240 optimisation runs in hours

FEM Model	Min	Max	Average	Total
Silastic	00:01:46	00:19:07	00:02:48	**02:52:04**
Wacker	00:01:27	00:27:55	00:03:10	**03:41:49**
Keyence-H	00:01:34	00:31:41	00:09:28	**11:13:48**
Keyence-L	00:02:19	00:34:44	00:06:58	**07:51:22**

5 Summary and Conclusion

The use of AM technologies for prototyping of aircraft door seals is presented, focusing on traditional silicone and UV-cured photopolymers. A mechanical characterisation was carried out with the help of uniaxial tensile experiments, along with a multiparametric design optimisation. The experiments reveal different kinds of mechanical behaviours depending on the distinct chemistry and polymerisation processes, which were taken into account for the design solutions with the genetic algorithm (NSGA-II).

In Sect. 1, the objectives of this work and the state of the art are described. Section 2 presents an overview of silicone AM technologies considered for the door seal prototype. Within this framework, Subsect. 2.2 portrays the experiments performed under uniaxial loading conditions. Section 3 presents the applied hyperelastic models and Drucker's stability. Subsection 3.2 compares the parameters of the four different silicones.

Subsequently, in Sect. 4, optimisation tasks are built for a prototype seal where part of the geometric variables are parametrised and constrained. Section 4.1 highlights the failure modes to be avoided in the context of an aircraft door seal, while also expressing the output variables obtained from the FEM analyses that should be minimised. Subsection 4.2 formalises the multiobjective minimisation problem and exemplifies how the genetic algorithm searches for optimal design candidates. In Subsect. 4.3, the results are summarised. By plotting the results of the deformed seal after the cabin pressure loading step, the majority of the design solutions were feasible, with the exception of the Keyence-L silicone, whose material stability was not met. Even though the "best" solutions are still associated to Keyence's technologies, a more consistent material modelling for these silicones could enable even better seal designs.

For future studies, it would be interesting to consider other material models that can better capture the material behaviour of certain elastomers, avoiding problems of polyconvexity. Furthermore, the modelling of the boundary problem was simplified in this study compared to real aircraft door loadings and should be more deeply explored. Relative deformations of the striker with respect to the nominal seal compression during ground and flight events, as well as temperature influence and time-dependent material behaviour should be incorporated to the optimisation guidelines of seal prototypes in order to mitigate failure scenarios.

6 Appendix

Table 7 presents the 3D printing parameters for the Silastic specimens:

Table 7. Printing parameters for Silastic (LAM)

Nozzle diameter	0.4mm
Layer Height	0.4mm
Line Width	0.41mm
Flow	145%
Retraction	10.0mm at 16.7mm/s
Outlines	0
Infill	100%
Speed	6.7mm/s

Acknowledgments. Parts of this research work are funded by dtec.bw - Digitalization and Technology Research Center of the Bundeswehr, which we gratefully acknowledge [project FLAB-3Dprint].

References

1. ACEO: 3D Printing With Silicones - Technology and Material. https://www.aceo3d.com/3d-printing/
2. Baturynska, I., Semeniuta, O., Martinsen, K.: Optimization of Process Parameters for Powder Bed Fusion Additive Manufacturing by Combination of Machine Learning and Finite Element Method: A Conceptual Framework. Procedia CIRP **67**, 227–232 (2018). http://dx.doi.org/10.1016/j.procir.2017.12.204
3. Bhandari, S., Lopez-Anido, R.: Finite element analysis of thermoplastic polymer extrusion 3D printed material for mechanical property prediction. Additive Manufacturing **22**(October 2017), 187–196 (2018). https://doi.org/10.1016/j.addma.2018.05.009
4. Bikas, H., Stavropoulos, P., Chryssolouris, G.: Additive manufacturing methods and modeling approaches: A critical review. International Journal of Advanced Manufacturing Technology **83**(1-4), 389–405 (2016). https://doi.org/10.1007/s00170-015-7576-2
5. Coogan, T.J., Kazmer, D.O.: Prediction of interlayer strength in material extrusion additive manufacturing. Additive Manufacturing **35**(May) (2020). https://doi.org/10.1016/j.addma.2020.101368
6. DIN Deutsches Institut für Normung: Prüfung von Kautschuk und Elastomeren - Bestimmung von Reißfestigkeit, Zugfestigkeit, Reißdehnung und Spannungswerten im Zugversuch (DIN 53504:2017-03) (2017)
7. Ding, J., Colegrove, P., Mehnen, J., Williams, S., Wang, F., Almeida, P.S.: A computationally efficient finite element model of wire and arc additive manufacture. International Journal of Advanced Manufacturing Technology **70**(1-4), 227–236 (2014). https://doi.org/10.1007/s00170-013-5261-x

8. Drucker, D.C.: A definition of stable inelastic material. American Society of Mechanical Engineers (1959)
9. Fancello, E.A.: Topology optimization for minimum mass design considering local failure constraints and contact boundary conditions. Structural and Multidisciplinary Optimization **32**(3), 229–240 (2006). https://doi.org/10.1007/s00158-006-0019-9
10. Ford, S., Despeisse, M.: Additive manufacturing and sustainability: an exploratory study of the advantages and challenges. Journal of Cleaner Production **137**, 1573–1587 (2016). https://doi.org/10.1016/j.jclepro.2016.04.150
11. Franke Goularte, B., Zatko, V., Lion, A., Johlitz, M.: Elastomeric door seal analysis under aircraft cabin pressure. Journal of Rubber Research **24**(2), 301–318 (2021). https://doi.org/10.1007/s42464-020-00071-7
12. Guessasma, S., Belhabib, S., Nouri, H., Ben Hassana, O.: Anisotropic damage inferred to 3D printed polymers using fused deposition modelling and subject to severe compression. European Polymer Journal **85**, 324–340 (2016). http://dx.doi.org/10.1016/j.eurpolymj.2016.10.030
13. Hartmann, S.: Numerical studies on the identification of the material parameters of Rivlin's hyperelasticity using tension-torsion tests. Acta Mechanica **148**, 129–155 (2001). https://doi.org/10.1007/BF01183674
14. Hartmann, S., Neff, P.: Polyconvexity of generalized polynomial-type hyperelastic strain energy functions for near-incompressibility. International Journal of Solids and Structures **40**(11), 2767–2791 (2003). https://doi.org/10.1016/S0020-7683(03)00086-6
15. Haupt, P.: Continuum Mechanics and Theory of Materials. Springer-Verlag, Berlin (2000)
16. Haupt, P., Lion, A.: A generalisation of the Mooney-Rivlin model to finite linear viscoelasticity. In: D. Besdo, R. Schuster, J. Ihlemann (eds.) Constitutive Models for Rubber II: Proceedings of the 2th European Conference on Constitutive Models for Rubber, pp. 57–64. Swets and Zeitlinger, London (2001)
17. Hibbitt, K.: ABAQUS/Standard: User's Manual: Version 5.6 (1996)
18. James, A.G., Green, A.: Strain energy functions of rubber. II. The characterization of filled vulcanizates. Journal of Applied Polymer Science **19**(8), 2319–2330 (1975). https://doi.org/10.1002/app.1975.070190822
19. Janusziewicz, R., Tumbleston, J.R., Quintanilla, A.L., Mecham, S.J., DeSimone, J.M.: Layerless fabrication with continuous liquid interface production. Proceedings of the National Academy of Sciences of the United States of America **113**(42), 11,703–11,708 (2016). https://doi.org/10.1073/pnas.1605271113
20. Joshi, S.C., Sheikh, A.A.: 3D printing in aerospace and its long-term sustainability. Virtual and Physical Prototyping **10**(4), 175–185 (2015). https://doi.org/10.1080/17452759.2015.1111519
21. Klahn, C., Leutenecker, B., Meboldt, M.: Design strategies for the process of additive manufacturing. Procedia CIRP **36**, 230–235 (2015). http://dx.doi.org/10.1016/j.procir.2015.01.082
22. Kumar, L.J.;Nair, C.: Current trends of additive manufacturing in the aerospace industry. In: Advances in 3D Printing and Additive Manufacturing Technologies, pp. 39–54 (2017). https://doi.org/10.1007/978-981-10-0812-2
23. Lejeunes, S.: Modélisation de structures lamifiées élastomère-métal à l'aide d'une méthode de réduction de modèles. Ph.D. thesis, Université de la Méditerranée-Aix-Marseille II (2006)

24. Lepoivre, A., Boyard, N., Levy, A., Sobotka, V.: Heat transfer and adhesion study for the FFF additive manufacturing process. Procedia Manufacturing **47**(2019), 948–955 (2020). https://doi.org/10.1016/j.promfg.2020.04.291
25. Lim, C.W.J., Le, K.Q., Lu, Q., Wong, C.H.: An Overview of 3-D Printing in Manufacturing, Aerospace, and Automotive Industries. IEEE Potentials **35**(4), 18–22 (2016). https://doi.org/10.1109/MPOT.2016.2540098
26. Liu, R., Wang, Z., Sparks, T., Liou, F., Newkirk, J.: Aerospace applications of laser additive manufacturing (2017). http://dx.doi.org/10.1016/B978-0-08-100433-3.00013-0
27. Michaleris, P.: Modeling metal deposition in heat transfer analyses of additive manufacturing processes. Finite Elements in Analysis and Design **86**, 51–60 (2014). http://dx.doi.org/10.1016/j.finel.2014.04.003
28. Nickels, L.: AM and aerospace: An ideal combination. Metal Powder Report **70**(6), 300–303 (2015). http://dx.doi.org/10.1016/j.mprp.2015.06.005
29. Nouri, H., Guessasma, S., Belhabib, S.: Structural imperfections in additive manufacturing perceived from the X-ray micro-tomography perspective. Journal of Materials Processing Technology **234**, 113–124 (2016). http://dx.doi.org/10.1016/j.jmatprotec.2016.03.019
30. Van der Velden, Alex and Koch, P.: Isight design optimization methodologies. ASM handbook **22**, 79 (2010). https://doi.org/10.31399/asm.hb.v22b.a0005505
31. Zhu, W., Wang, J., Lin, P.: Numerical analysis and optimal design for new automotive door sealing with variable cross-section. Finite Elements in Analysis and Design **91**, 115–126 (2014). https://doi.org/10.1016/j.finel.2014.06.012

Development, Industrialization and Qualification of a Lever-Shaft-Integration for a Long Range Aircraft

Christian Wolf[1](✉), Andreas Neumann[2], and Sebastian Reinspach[1]

[1] Airbus Helicopters Deutschland GmbH, ETIET, Donauwörth, Germany
{christian.chri.wolf,sebastian.reinspach}@airbus.com
[2] Airbus Helicopters Deutschland GmbH, ETIES, Donauwörth, Germany
andreas.a.neumann@airbus.com

Abstract. Lightweight design is a powerful method to decrease weight, to increase the performance of an application. This document describes the design, the development (industrialization), the qualification of a lightweight design application (airplane door latch) which was developed under application of additive manufacturing (Ti6Al4V-powder-bed-fusion). The document explains the development content, the requirements, the applied design, development considerations, the sizing operation. A multi-process customized design solution, based on an integrated design, represents the key innovation. Furthermore the paper explains the qualification, certification approach, provides an example for the achieved mechanical properties. The final result of the development is a decrease in weight by 45%, at once a decrease in recurring cost by 25%. The additively manufactured airplane door latch received certification by EASA, FAA. It is an example of the potential of additive manufacturing in an aerospace environment for existing, new aircrafts.

Keywords: Additive manufacturing · Lever-shaft-integration · EASA/FAA certified

1 Preface

Airbus Helicopters offers the full spectrum of rotary wing aircraft solutions to its customers in more than 150 countries worldwide [1]. In parallel Airbus Helicopters is a design and build partner for Airbus Commercial Aircraft in the area of cabin entry systems for large fixed wing aircrafts such as passenger entry doors, cargo doors, bulk cargo doors and emergency exit doors. In the year 2019 this complementary business of Airbus Helicopters employed a work force of 1350 employees and delivered 5000 door units to Airbus. 'Airplane Doors' supplies approximately 70% of the cabin-entry needs for the Airbus fixed wing aircraft production. It manufactures conventionally fabricated, aluminum-riveted doors as well as doors equipped with a CFRP structure.

J. Rieser et al. (Eds.): *Proceedings of the Munich Symposium on Lightweight Design 2021*, Proceedings, pp. 154–159, 2023. https://doi.org/10.1007/978-3-662-65216-9_14

In the year 2016 Airbus Helicopters presented the second generation of a CFRP passenger entry door. The door system conceived new CFRP load introduction solutions (ie the maxShear load introduction) which make metallic reinforcements obsolete. At the same time Airbus Helicopters and its partner Fraunhofer LBF, Darmstadt introduced a new door test facility. The door test system offers 50 DOF[1] for a full compliant representation of an aircraft fuselage door interface. The test facility allows a double-cabin pressure load and measures the reaction forces via 3D load cells. Both innovations together emphasis the research and product development capabilities of 'Airbus Helicopters – Airplane Doors' and the role as a market leading design and build partner for Airbus.

2 Introduction

State-of-the-art airplane doors fulfill the definition of a system. The system elements are a door structure, an opening and closing mechanism, a main hinge, an electric status monitoring, a thermo-acoustic insulation and a cabin lining. The doors main functions are the provision of fuselage functions in cruise condition, the provision of access to the cabin for passengers and crew members and the provision of the emergency exit function.

The door opening motion is initially inward, upward and a swivel motion outward. Aircraft doors for large passenger aircrafts follow EASA/FAA certification specification 25 (CS-25) [2]. The certification specification does not accommodate a doors chapter. Never-the-less there are some door specific requirements such as 'multiple layers of protection' for passenger doors. A history of accidents in the 1960's and 1970's with airplane doors led to the introduction of additional safety elements. The multiple layers of protection consist of a latching system, a locking system, an indication system and a pressure prevention means. The basic protection objective is to provide a high degree of tolerance to failures, malfunction and human errors. The CS-25 paragraph illustrates the significance of the door latch to the safety of the aircraft. From an 'aircraft-system' point of view the door latch represents a flight critical (important) element of the aircraft.

3 Requirements of the Door Latch and Design Considerations

The research project GENfly [3] generated many promising results with respect to the material quality of additively manufactured components. Based on these positive results Airbus Helicopters decided to perform research and pre-development activities with the objective to improve the weight and the cost of the A350XWB passenger door latching system. The pre-development work was built-on the components requirements

[1] DOF: degree of freedom.

such as material requirements, load requirements, installation and door mechanism requirements, process requirements and quality assurance requirements. It is important to mention that the door latch is loaded mainly by static forces. Its cyclic loads are very small compared to the static loads which is an advantageous situation for an additive manufacturing application. Material requirements are defined in several CS-25 paragraphs and subsequent in the relevant Airbus company standards. An important requirement for additively manufactured components is the surface quality. Oxygen modifications such as alpha-case or an increase in micro-hardness in the fringe of the part are not allowed. Additionally the Airbus material standards require fluorescence penetrant inspection (FPI) together with radiographic testing such as digital x-ray or computerized tomography. Furthermore the geometrical compliance, and the way of demonstrating this, are important requirements for the development (Fig. 1).

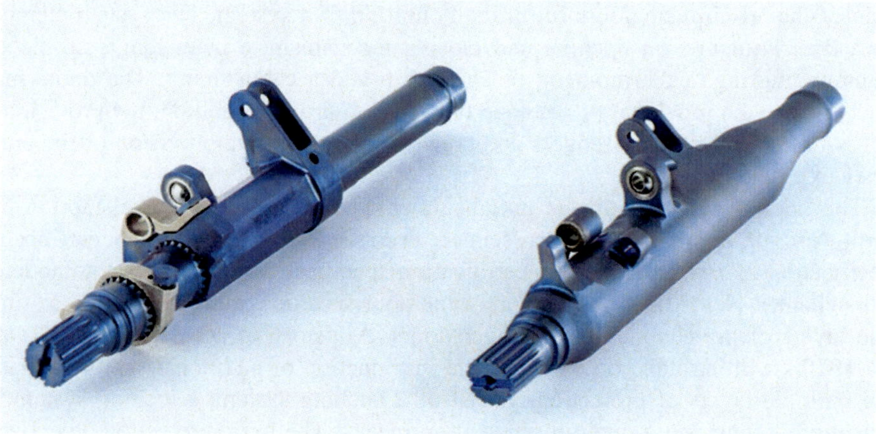

Fig. 1. Left hand side: The conventionally latch shaft assembly consists of 10 parts. Right hand side: the newly developed *AM latch shaft* represents a single part which integrates ten parts into one part

4 Mechanical System and Sizing Operation

The development of the *AM latch shaft* was based on a systematic design approach. Under application of the requirements different solutions in the development phases 'physical principle' and 'concept' came into existence.

Iteration 1 demonstrated a truss-structure combined with local reinforcements and strongly pronounced bionic elements. After the creation of early demonstrators the design developed into a closed surface structure including bionic elements (iteration 2). Iteration 3 applies the surface structure again and includes an optimized topology. The concept of integration is common to all design iterations and represents the most powerful solution method. The *AM latch shaft* realizes an integration ratio of ten parts into one part. The integrated design (iteration 3) of the *AM latch shaft* eliminates overlap areas known from conventionally manufactured assemblies. In addition to that it presents stress-customized wall thicknesses and manufacturing friendly geometries.

The *AM latch shaft* is based on a cantilever beam. Due to the latch lever which is assembled to the main body a combined loading by torsion and bending comes into existence (door operation). The critical load case is slightly different compared to the standard operation but still torsion and bending are the most important stress components. As already described the application uses stress-customized wall thicknesses to react the applied loads.

5 Multi-Process Optimization

Lightweight design represents a multi-functional field of science. Lightweight design principles and state-of-the-art sizing methods are very important but lightweight design strategies also includes manufacturing process elements such as production light weight design [4]. The Airbus Helicopters additive manufacturing production system integrates the full additive value chain. It starts with the data preparation (print-pre-stage), the print, de-powdering, stress relief heat treatment and the separation from the build plate. This step of master forming is followed by sand blasting, hot-isostatic-pressing and etching. After this the functional surfaces and fittings are created by machining. The final sequence includes computerized tomography, fluorescence penetrant inspection (FPI) and the installation of bushes and bearings. It is important to understand that the print process itself consumes 30% of the recurring cost. 70% of the recurring cost are created by the post print operations. The development activities at Airbus Helicopters are based on a commercial environment and are limited by schedule and budget constraints. In order to comply with the commercial environment Airbus Helicopters applies a multi-optimization development method. This means the activities of geometry creation, sizing, the development of the print job, the development of the machining operations and the AM process definition are performed simultaneously. The concurrent approach allows the development team to introduce new solution elements and to immediately perform manufacturing trials to instantly validate the new features before freezing the final configuration. This approach requires precise coordination but allows to develop a process customized lightweight design application within a very short cycle time compared to a sequential operation.

6 Certification

The certification of the *AM latch shaft* applies the basic concept of part qualification (validation of the application performance). The defined process (frozen manufacturing route), the design and the Airbus internal additive manufacturing (AM) standards constitute the certification basis. The assessment of the application performance includes the validation of the Material-Parameter-Supplier (printing infrastructure) set-up, the validation of the part characteristics and a destructive component test. The validation of the Material-Parameter-Supplier (printer infrastructure) set-up is basically a test campaign to evaluate the mechanical properties of the melted material.

The validation of the part characteristics includes the geometrical compliance, the microstructure of the material, porosity and lack-of-fusion inspections as well as FPI inspections to exclude surface defects. The destructive component test represents a cross-check of all other performed validation methods and ensures the overall mechanical performance of the AM application.

The modification justification report delivered a number of very advantageous results. The fatigue strength of the Airbus Ti6Al4V-material surpasses the specification limits by 5–10%. The *AM latch shaft* represents the replacement of an existing design solution. In order to safeguard the door system qualification the Airbus Helicopters development team targeted on an equal performance compared to the previous solution. The component test (static strength) resulted in a slight increase of the stiffness. With respect to the part strength a sufficient reserve factor (compared to the strength requirement) could be achieved (Fig. 2).

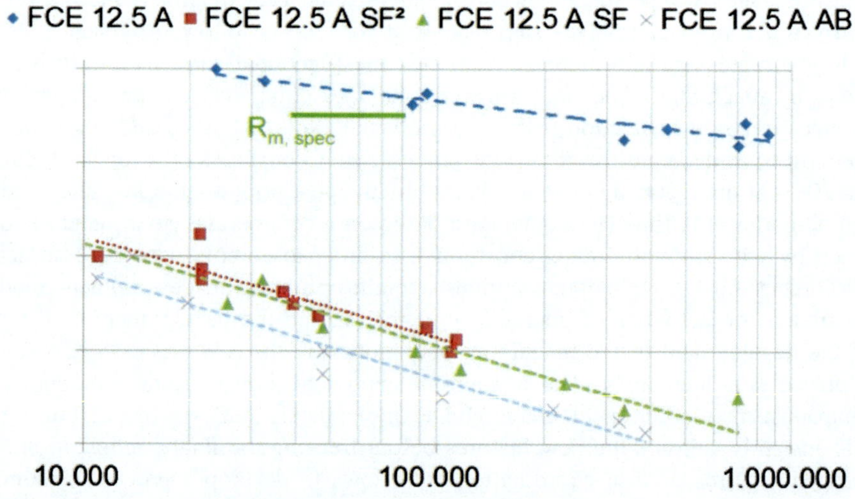

Fig. 2. Qualification test result of the fatigue strength of Ti6Al4V produced by means of additive manufacturing in the framework of the qualification and certification activities of the *AM latch shaft*. The graph compares test results in 'as-build' surface quality (AB) with specimen in 'surface finished' quality (SF), specimen in 'two-times surfaced finished' quality (SF^2) and 'machined' specimen (no suffix)

7 Achievements

After a successful validation test campaign (including the characterization of the melted material, the evaluation of the part characteristics and a series of component tests) Airbus and Airbus Helicopters decided to apply the *AM latch shaft* for the serial manufacturing of the A350XWB passenger doors. The development resulted with a decrease in weight by 45% and at once with an increase in the parts performance. This performance increase is well appreciated since the latch shaft is flight critical. At the same time the new additively manufactured solution decreases the recurring cost by 25%. Later process optimizations generated a potential to decrease the application weight to 50% at a 50% recurring cost reduction. The development objective to generate a commercial advantage together with a significant weight reduction could be achieved. Both certification organizations EASA and FAA acknowledged the development and certified the application. The EASA Major Change Approval 10,074,979 was awarded 26 Nov 2020. The confirmation of FAA has been communicated 08 Sep 2021. Between the start of production and today Airbus Helicopters manufactured 3200 parts and replaced a number 32.000 conventionally manufactured parts. The AM latch shaft represents a flight critical part fabricated by additive manufacturing and is certified by the airworthiness authorities. It is the first of such parts in civil aviation history. This achievement represents a major milestone with respect to the commercial application of additive manufacturing for Airbus Helicopters as well as for Airbus Group.

References

1. Airbus Helicopters homepage, http://www.airbus.com/helicopters.html, last accessed 2021/10/31.
2. EASA homepage, https://www.easa.europa.eu/document-library/certification-specifications/cs-25-amendment-26-0, last accessed 2021/10/31.
3. TIB Leibnitz Informationszentrum Technik und Naturwissenschaften, Universitätsbibliothek homepage: https://www.tib.eu/de/suchen?tx_tibsearch_search%5Bquery%5D=genfly&tx_tibsearch_search%5Bsrt%5D=rank&tx_tibsearch_search%5Bcnt%5D=20, last accessed 2021/10/31.
4. Henning, F., Moeller E., (ed.): Leichtbauhandbuch, Methoden, Werkstoffe, Fertigung. 1st edition, Carl Hanser, München, Wien (2011).

Characterization and Influences of the Load Carrying Capacity of Lightweight Hub Designs of 3D-Printed Gears (16MnCr5, PBF-LB/M-Process)

Karl Jakob Winkler[1](✉), Matthias Schmitt[2], Thomas Tobie[1], Georg Schlick[2], Karsten Stahl[1], and Rüdiger Daub[2,3]

[1] Gear Research Center (FZG), Technical University of Munich (TUM), Garching, Deutschland
{karljakob.winkler,thomas.tobie,karsten.stahl}@tum.de
[2] Fraunhofer Institute for Casting, Composite and Processing Technology IGCV, Augsburg, Deutschland
{matthias.schmitt,georg.schlick,}@igcv.fraunhofer.de, ruediger.daub@iwb.tum.de
[3] Institute for Machine Tools and Industrial Management (iwb), Technical University of Munich (TUM), Garching, Deutschland

Abstract. Lightweight design for gears is becoming increasingly important for efficient, sustainable drive trains. Innovative lightweight designs can be achieved by the additive manufacturing process of P.B.F.-L.B./M. (powder bed fusion by laser beam of metals). This contribution presents lightweight hub designs for gears manufactured by P.B.F.-L.B./M. Helical as well as spur gears are 3D-printed out of the case-hardening steel 16MnCr5. After the PBF-LB/M-process, the gears are case-carburized, shot blasted for mechanical cleaning. The gears with lightweight hubs are analyzed concerning their density, microstructure, roughness. The gears are tested regarding their static, dynamic load carrying capacity, the influence of the lightweight hub on the load carrying capacity is analyzed, evaluated. In conclusion, this contribution enables aprofound understanding, prospective evolution of lightweight hub designs for gears.

Keywords: Load carrying capacity · Lightweight · Gear · Hub · 3D-printed · 16MnCr5 · PBF-LB/M · LPBF

1 Introduction

Within the introduction, the PBF-LB/M process will be introduced as well as the application of the PBF-LB/M process on the manufacturing of gears and the difference to conventional manufacturing of gears.

J. Rieser et al. (Eds.): *Proceedings of the Munich Symposium on Lightweight Design 2021*, Proceedings, pp. 160–174, 2023. https://doi.org/10.1007/978-3-662-65216-9_15

1.1 PBF-LB/M Process (LPBF)

The term PBF-LB/M is an acronym, stands for powder bed fusion (PBF) with a laser beam (LB) of metals (M) and is an additive manufacturing process. PBF-LB/M is also known as LPBF, which stands for laser powder bed fusion or the non ISO-term "selective laser melting (SLM)". PBF-LB/M is an additive manufacturing process enabling new possibilities with regard to design and manufacturing. Conventional manufacturing restraints can be reduced and innovative designs are achievable. Conventional manufacturing techniques rely on primary forming a gear by casting and forging as well as removing material by cutting, milling and grinding. Using PBF-LB/M it is possible to create undercuts and curved bores, both of which are not easily realized with conventional manufacturing. With the new design freedom of additive manufacturing, functionality features such as curved cooling pipes or mass reduction by lightweight designs are possible. The realization of such functionality and mass reduction has been proven by the authors in recent research and publications. [1–3]

Additive manufacturing is characterized by a layer-by-layer build-up of the geometry [4, 5]. PBF-LB/M is a repetitive process, with the following steps:

1. Powder is distributed by the coater from the powder supply onto the building platform. Additional powder is discarded in the powder throw-out.
2. Powder is melted by a controlled laser beam according to the component geometry. Powder will be solidified, where needed for the component geometry, the rest of the powder remains unsolidified.
3. The building platform is lowered, the powder supply is raised and the process repeats itself.

The entire process is performed under a shielding gas atmosphere. Common shielding gases are nitrogen, argon or helium. For the research in this paper, argon is used as a shielding gas. Figure 1 shows a schematic process set-up of powder bed fusion with a laser beam of metals (PBF-LB/M).

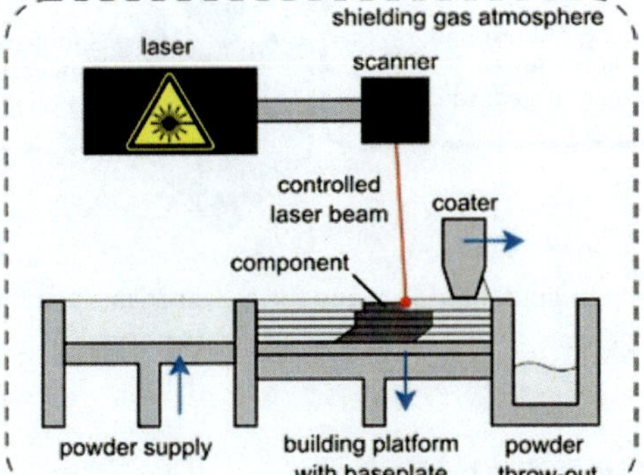

Fig. 1. Schematic process set-up of powder bed fusion with a laser beam of metals (PBF-LB/M)

1.2 Gears Manufactured by PBF-LB/M

The gears investigated in this paper are manufactured by the PBF-LB/M process. Gears are a widespread and usually highly-loaded component in drive trains and gear boxes. In order to raise the material strength, these parts are typically made out of case-hardening steels such as 16MnCr5 and hardened during a heat-treatment called case-carburizing after soft machining of the gear geometry. Besides the heat treatment, primary forming, soft machining and grinding are important manufacturing steps of conventional manufactured gears. As a pre-processing, the powder production by e.g. gas atomization for the PBF-LB/M process as well as the smeltering for the conventional process have to be realized. The main difference between conventional and additive manufacturing is the primary forming by PBF-LB/M instead of casting and forging. Due to the different primary forming, the subsequent soft machining is different as well. The heat treatment and the hard machining (grinding) for conventional and PBF-LB/M manufacturing are comparable. Figure 2 shows a comparison of the manufacturing steps for a gear manufactured conventionally and by PBF-LB/M.

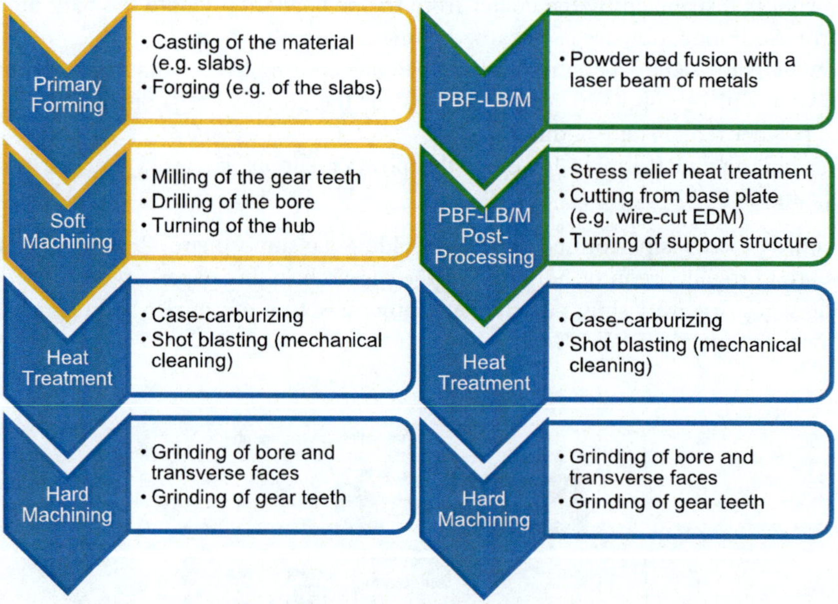

Fig. 2. Comparison of the manufacturing steps for a gear manufactured conventionally and by PBF-LB/M (exemplary process route, alternative steps are possible)

2 Lightweight Gear Hub Designs

Two major lightweight gear hub designs were developed and investigated within the research of this paper: a topologically optimized and a bionically optimized gear hub design. The bionically optimized gear hub design was elaborated into two versions:

a (conservative) bionically and an advanced bionically optimized gear hub design. Besides the lightweight gear hub designs, conventional full material hubs were additively manufactured for comparison.

The process of developing and optimizing the lightweight gear hub designs is comprehensively described in the publication "Framework and strategies for the lightweight construction of AM gears for the automotive industry" [4]. The main information regarding the lightweight gear hub designs is summarized in the following chapters to better understand the later following investigations on the characteristics and the influence on the load carrying capacity.

2.1 Topologically Optimized Gear Hub Design

The topologically optimized gear hub design is developed with the help of the simulation programs Inspire 2018 and Siemens NX 11. The design space is defined by a gear rim thickness and a minimal thickness above the feather key. The gear rim thickness is based on calculations according to ISO 6336 [6]. The minimal thickness above the feather key is based on experience of the gear research institute (FZG) and scientific publications [7]. The gear models are positioned, paired and meshed before simulations with a static load are performed. Figure 3 shows a visualization of the design space as well as the positioning of the gears, a detail view of the pairing and the meshing refinement.

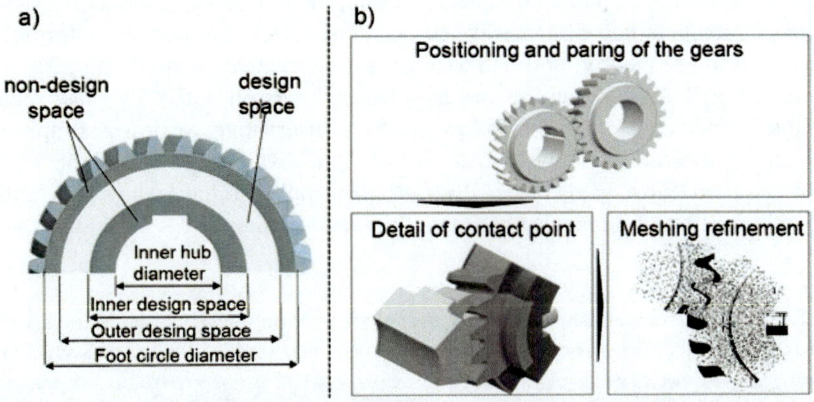

Fig. 3. Visualization of **a** the design space and **b** the positioning of the gears, a detail view of the pairing and the meshing refinement [4]

For the topologically optimization of the gear hub design, a variation of the symmetry planes is performed. The initial number of symmetry planes is 27: one symmetry plane for each tooth of the gear. With each step of the variation the number of symmetry planes is reduced by 40% until six symmetry planes are reached. Based on a FEM simulation, the variant with 16 symmetry planes is chosen due to an even load distribution. Figure 4 shows a visualization of the variation of the symmetry planes.

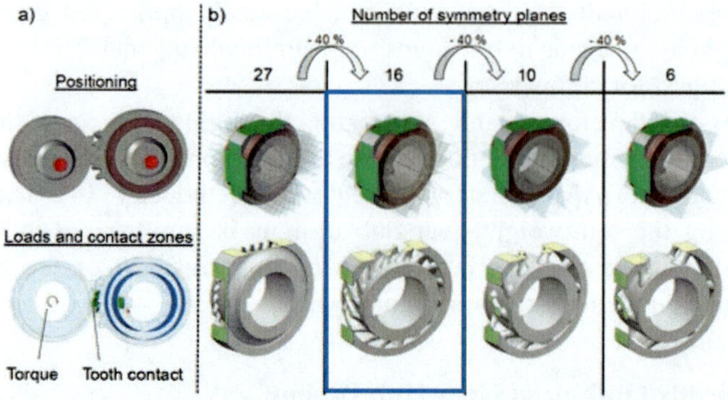

Fig. 4. Visualization of **a** the positioning, loads and contact zones and **b** the variation of the symmetry planes in the topology optimization [4]; variant with 16 symmetry planes is chosen

2.2 Bionically Optimized Gear Hub Design

The bionically optimized gear hub design is developed with the help of the simulation program Siemens NX 11. The design space is identical to the topologically optimized gear hub designs. Three biological models are selected and their structure abstracted and transferred to a gear hub design. All gear hub designs were roughly analyzed with the help of FEM to make sure, that a basic strength of the lightweight gear hubs is ensured. Based on the calculated mass reduction, the bulrush as a model for the bionically optimized gear hub design was selected. The selected bionically optimized gear hub design is subsequently modified to a more conservative version with 30% instead of the initial 35% mass reduction and an advanced version with 45% mass reduction (see Table 2 of Sect. 2.3). In the following, the conservative version is simply called "bionically optimized hub" without the label "conservative", whereas the advanced version is called "advanced, bionically optimized hub". Table 1 shows the different biological models and bionically optimized gear hub designs.

Table 1. Biological models and bionically optimized gear hub designs; the bulrush variant is selected and modified to a conservative and an advanced version (see Table 2 of Sect. 2.3)

Designs	Concept 1 Winter Horsetail	Concept 2 Blue Alga	Concept 3 Bulrush	Concept 3 Advanced
Biological Model				
Bionic Gear Design	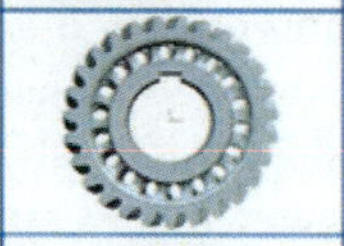	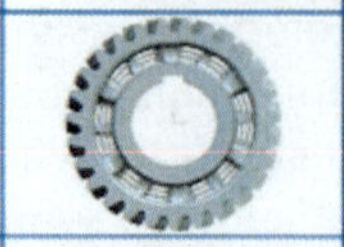	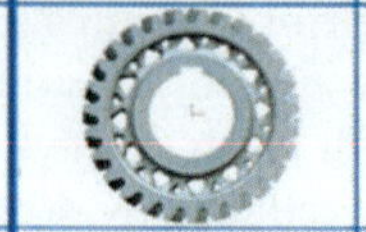	
Mass reduction	32 %	32 %	35 %	45 %
Safety at 250 Nm / 1450 N/mm²	1,18	0,94	0,96	0,96

2.3 Manufactured Gears and Lightweight Hub Designs

There are three types of gears and three main types of lightweight hub designs manufactured. The gear types are common test gear geometries which have been in use at the gear research instute (FZG) and for which extensive experience is available. The manufactured gears and their corresponding hub designs are listed in the following:

- Type 1 gears (helical, $m_n = 3.3$ mm, $z_{1/2} = 25/27$) with:
  - conventional, full material hubs
  - bionically optimized hubs
  - topologically optimized hubs
  - advanced, bionically optimized hubs
- Type 17/18 gears (spur, $m_n = 5$ mm, $z_{1/2} = 17/18$) with:
  - conventional, full material hubs
  - two gear wheels with bionically optimized hubs based on the blue alga model (trial tests)
- Type Pm5 gears (spur, $m_n = 5$ mm, $z = 24$) with:
  - conventional, full material hubs
  - advanced, bionically optimized hubs

The hubs for the Type 1 and Type Pm5 gears are equally manufactured regarding quantity. The two gear wheels with bionically optimized hubs based on the blue alga model were trial tests to proof the concept of biological models apart from the bulrush.

The gears were tested regarding the following load carrying capacities:

- Type 1
  - static form stability
  - dynamic surface durability
- Type 17/18
  - dynamic surface durability
- Type Pm5
  - dynamic tooth root bending strength

The PBF-LB/M process creates a high surface roughness of up to $Rz = 60\ \mu m$. In order to reduce the surface roughness and to investigate its influence on the tooth root bending strength, half of the Type Pm5 gears with conventional, full material hubs and half of the Type Pm5 gears with advanced, bionically optimized hubs were chemically polished. The chemical polishing consisted of a bath in acid for a defined period of time. The exact characteristic values of the chemical polishing were not disclosed by the contract manufacturer. All transverse surfaces of the Type Pm5 gears were grinded in order to achieve a precise tooth width, which is essential for the experimental testing of the tooth root bending strength. The Type 1 and Type 17/18 gears were not chemically polished but tooth flank grinded for experimental investigations regarding the surface durability. Figure 5 shows the gear types with their corresponding lightweight hub designs and manufacturing routes.

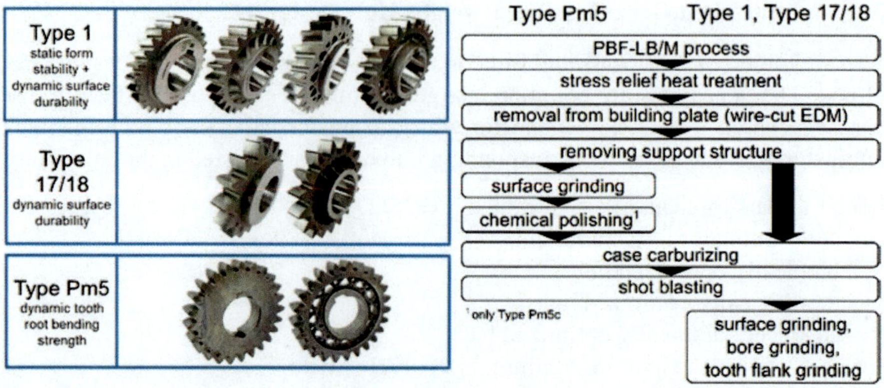

Fig. 5. Gear types with their corresponding lightweight hub designs and manufacturing routes

Table 2 shows the manufactured lightweight hub designs with their corresponding mass reduction for the Type 1 gears.

Table 2. Manufactured lightweight hub designs with their corresponding mass reduction for the Type 1 gears

Name	Conventional (full material hub)	Bionically optimized hub	Topologically optimized hub	Advanced bionically optimized hub
Design				
Mass reduction	[–]	25 %	30 %	45 %

3 Characteristics of Lightweight Gear Hub Designs

Regarding the characteristics of the lightweight gear hub designs the following properties were investigated: density, microstructure and hardness depth profile.

3.1 Density

The density was evaluated with the principle of Archimedes as well as with computer aided analysis of unetched gear sections. The computer aided analysis identified scarce and small-sized cavities within the gear sections, which are shown as minuscule black dots on the gear sections. Both methods resulted in relative densities of $\rho_{rel}>99{,}7\%$ compared to a material with no cavities ($\rho_{rel}=100\%$). The slight porosities were not agglomerated but evenly distributed over the entire gear section. The density did not show an influence on the load carrying capacity of the lightweight hub designs. No lightweight hub design failed due to its density. Figure 6 shows the cross sections for the density analysis of the gears with different lightweight hub designs.

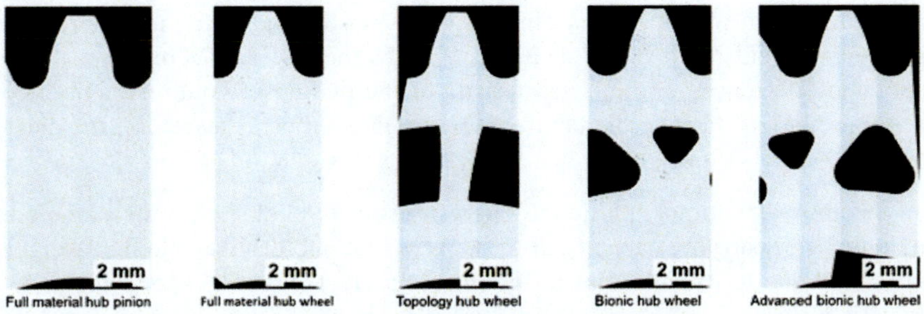

Fig. 6. Cross sections for density analysis of the different lightweight gear hub designs; cavities are shown as minuscule black dots on the gear sections

3.2 Microstructure

The microstructure of the different lightweight hub designs was analyzed with the help of etched metallographic gear sections. The different colors in the cross sections are caused by small time deviations regarding the etching and can be neglected. All cross sections show a constant case-hardening depth and a homogenous case-hardening layer. The case-hardening layer is made up of a mainly martensitic microstructure, the material beyond the case layer is a composition of mainly upper and lower bainitic microstructures. The lightweight structures were not through-hardened, which was explicitly intended to avoid a brittle and thus fragile microstructure. In general, the microstructure of the additively manufactured gears is comparable to conventional, smelt-metallurgical gears. Figure 7 shows the etched cross sections of the lightweight hub designs for the analysis of the microstructure.

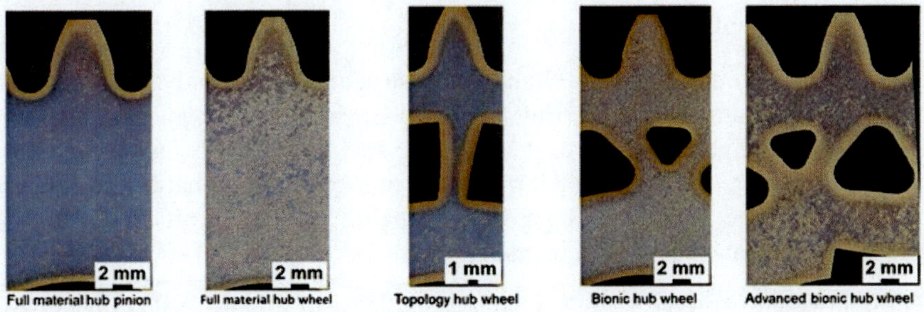

Fig. 7. Etched cross sections of the lightweight gear hub designs for analysis of the microstructure (different colors due to small deviations of the etching time can be neglected)

Apart of the generally comparable microstructure, the following particular characteristics regarding the microstructure of PBF-LB/M gears were found:

- Scarce and evenly distributed foreign material occurrences (especially aluminum-oxides) possibly due to powder cross-contamination (handling and processing of powder on a research PBF-LB/M machine that is used with different types of powder).
- Pronounced non-martensitic case layers with troostite and ferrite microstructures.
- Scarce and evenly distributed cavities caused by the PBF-LB/M process.
- Sporadic differences in the characteristic of the martensitic microstructure: coarse and fine martensitic microstructures next to each other; research on the causes is ongoing.

All of the above-mentioned particular characteristics did not show a measurable influence on the load carrying capacity of the lightweight hub designs. No lightweight hub design failed due to irregularities in the microstructure. Figure 8 shows the particular characteristics regarding the microstructure of lightweight hub designs manufactured by PBF-LB/M.

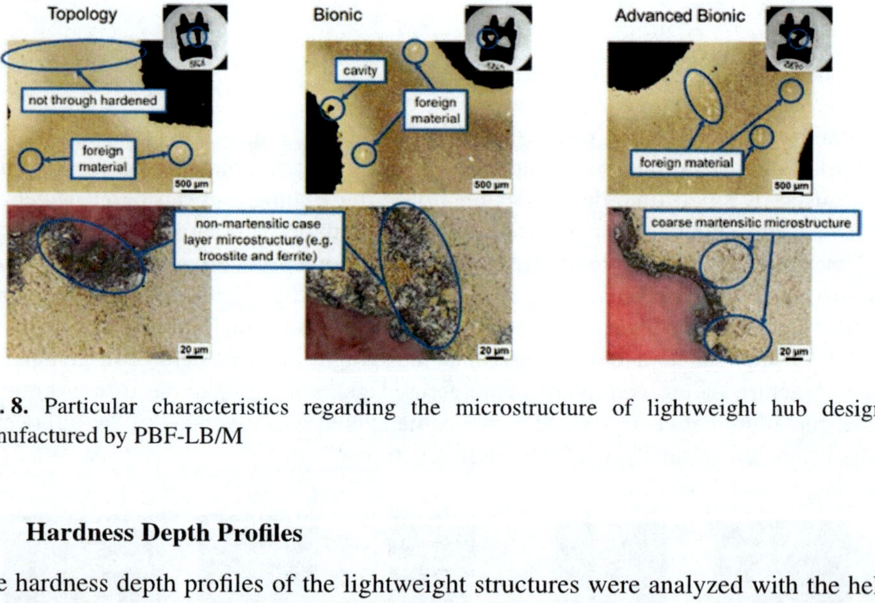

Fig. 8. Particular characteristics regarding the microstructure of lightweight hub designs manufactured by PBF-LB/M

3.3 Hardness Depth Profiles

The hardness depth profiles of the lightweight structures were analyzed with the help of gear sections and a hardness testing machine. The hardness was measured from one side of the lightweight structure to the opposite side of the lightweight structure. All lightweight structures are at least 3.0 mm thick to avoid through-hardening and brittle microstructures (see Sect. 3.2). Regarding the hardness depth profiles, the following observations and conclusions can be made:

- All lightweight structures show a surface hardness of around 700 HV1 and a core hardness of around 300 HV1. The core hardness was measured at the intersection of normals constructed on the points of contact of the 30°-tangent to the tooth root fillet, the position is basically in the gear rim underneath a tooth. These values are comparable to conventional, case-carburized gears and lie within the material quality MQ specified in part 5 of ISO 6336 [6].

- All lightweight structures show a symmetrical hardness depth profile, which correlates with a homogenous heat treatment.
- All hardness values in the center of the lightweight structures are above the core hardness, which correlates with the higher cooling rates of the lightweight structures compared to the thicker gear rim.
- The tendency towards higher hardness values in the middle of the lightweight structure correlates with the mass reduction of the lightweight gear hub design: the less material, the higher the hardness values. This observation correlates with the knowledge, that less material leads to higher cooling rates which leads to higher amounts of martensitic microstructures.

All of the above-mentioned characteristics did not show a measurable influence on the load carrying capacity of the lightweight hub designs. No lightweight hub design failed due to its hardness depth profile. In sum, the hardness depth profiles correlate well with the expectations derived from conventional hardness depth profiles. Figure 9 shows the hardness depth profiles of the lightweight structures.

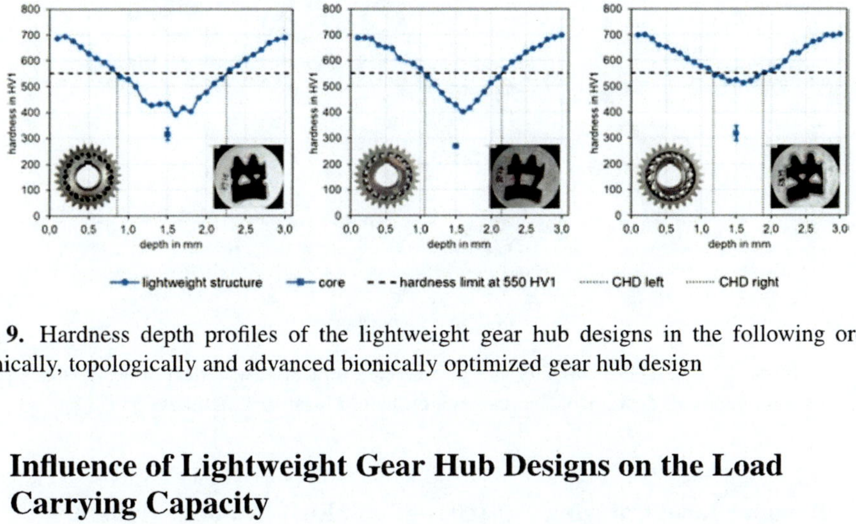

Fig. 9. Hardness depth profiles of the lightweight gear hub designs in the following order: bionically, topologically and advanced bionically optimized gear hub design

4 Influence of Lightweight Gear Hub Designs on the Load Carrying Capacity

The influence of the lightweight gear hub designs on the load carrying capacity was elaborated with the help of static and dynamic experimental investigations. A comprehensive description of the test rig, the experimental procedure and the analysis of the results has to be omitted due to the limited scope of this paper but can be found in the literature [8–10].

4.1 Static Load Carrying Capacity

The static load carrying capacity is tested with a static test rig for all three lightweight gear hub designs. Lightweight gears of the research project "Massiver Leichtbau" [11] are compared with the additively manufactured lightweight gears. The lightweight gears of the research project "Massiver Leichtbau" have a case-carburized gear rim which is joined with different lightweight hubs. Regarding the static load carrying capacity, the following observations and conclusions can be made:

- All lightweight hub designs endured the maximum torque of $T_{max} = 1200$ Nm structurally stable and without plastic deformations.
- Compared to the lightweight hubs of the research project "Massiver Leichtbau", the additively manufactured lightweight hub designs show a great potential regarding mass reduction and static load carrying capacity.

Figure 10 shows the static load carrying capacity of Type 1 gears with lightweight hub designs compared to lightweight gear hub designs of the research project "Massiver Leichtbau".

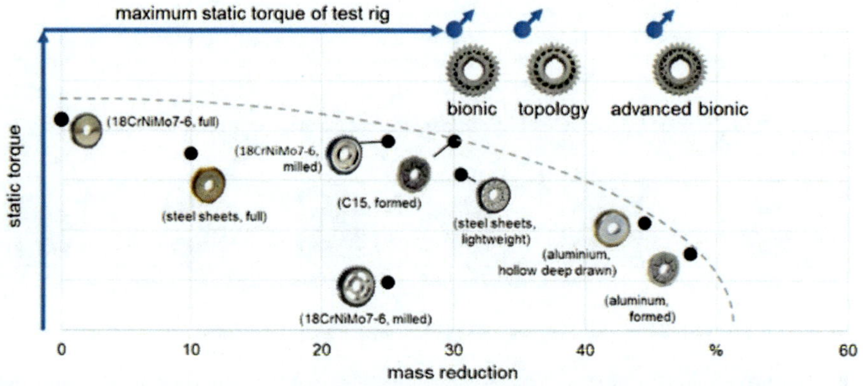

Fig. 10. Static load carrying capacity of Type 1 gears with lightweight hub designs compared to lightweight gear hub designs of the research project "Massiver Leichtbau" [11]

4.2 Dynamic Load Carrying Capacity—Tooth Root Bending Strength

The tooth root bending strength is evaluated with the help of a pulsator test rig. The four variants of Type Pm5 gear wheels differ in regard of the lightweight hub design and the chemical polishing. Based on the experimental investigations regarding the tooth root bending strength, the following observations and conclusions can be made:

- All lightweight gear hub designs were structurally stable and no plastic deformations were observed during the testing of the tooth root bending strength.
- Considering the common scattering of the experimental results, the chemically polished and the unpolished variants are comparable regarding the tooth root bending

strength. Additionally, the lightweight gear hub designs do not influence the tooth root bending strength.

- The chemical polishing does not deteriorate the structural stability of the lightweight hub designs.

In sum, the investigated lightweight gear hub designs did not show any influence on the tooth root bending strength. The structural stability during the tooth root bending tests shows the further potential regarding mass reduction by lightweight gear hub designs. Figure 11 shows the tooth root bending strength of the different variants of Type Pm5 gears with a reference from literature [12] and a comparison value from research [13]. The lower tooth root bending strength of the PBF-LB/M gears is due to the much higher surface roughness compared to the conventional reference values and will be considered more detailed in upcoming publications.

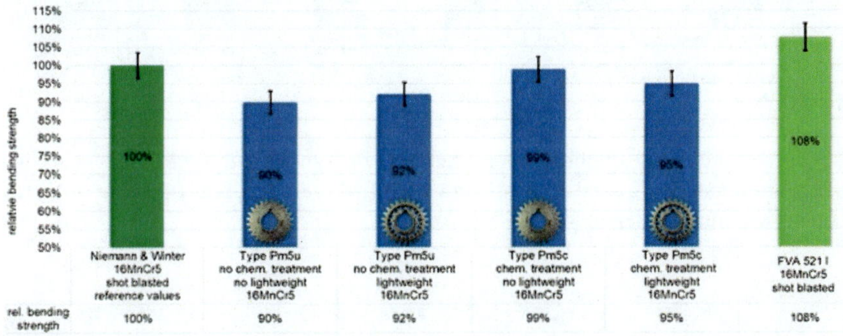

Fig. 11. Tooth root bending strength of the different variants of Type Pm5 gears with a reference from literature [12] and a comparison value from research [13] (*blue*: PBF-LB/M; *green*: smelt-metallurgical reference)

4.3 Dynamic Load Carrying Capacity—Surface Durability

The surface durability of the Type 1 and Type 17/18 gears was tested on a FZG back-to-back test rig. Based on the experimental investigations regarding the surface durability, the following observations and conclusions can be made:

- All but one lightweight gear hub designs were structurally stable and no plastic deformations were observed.
- One exception occurred with a structural failure of a Type 17/18 gear with a lightweight structure based on the blue alga. This failure was provoked by a known small material thickness above the feather key. The notch effect at the keyway most likely caused a stress peak which could not be sustained by the limited amount of material. It is strongly assumed, that the failure initiated at the feather key and not within the lightweight structure. Figure 13 shows the structural failure.
- Considering the common scattering of the experimental results, all variants are comparable regarding their surface durability. Within the Type 1 variants, load cycle numbers on a certain load level with different lightweight gear hub designs

are comparable. Thus, the investigated lightweight gear hub designs do not influence the surface durability.

Figure 12 shows the surface durability of the different variants of Type 1 and Type 17/18 gears with a reference from literature [12] and comparison values from research [13, 14].

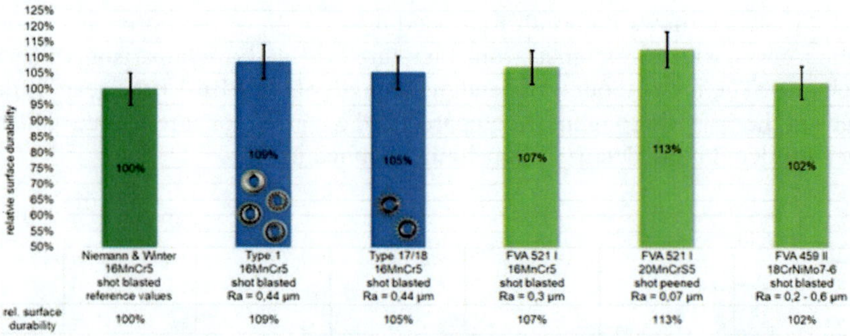

Fig. 12. Surface durability of the different variants of Type 1 and Type 17/18 gears with a reference from literature [12] and comparison values from research [13, 14] (*blue*: PBF-LB/M; *green*: smelt-metallurgical reference)

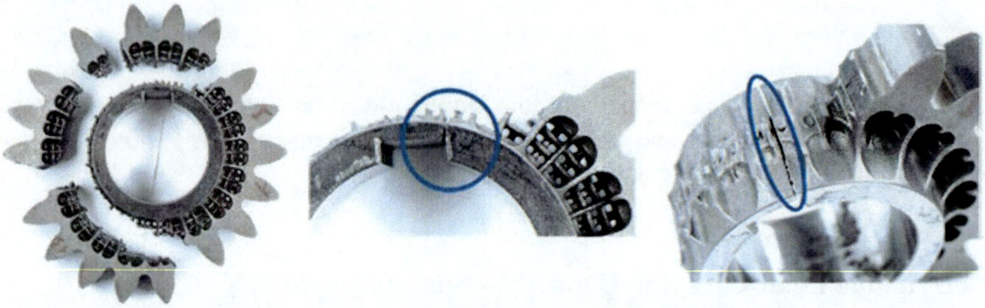

Fig. 13. Only structural failure of a test gear with lightweight gear hub design due to a limited material thickness above the feather key; failure most likely initiated by notch effect of the feather key, not within the structure of the lightweight hub design

5 Summary and Outlook

The summary of this paper includes the following main findings:

- Gears with topologically and bionically optimized lightweight hub designs can be created with the additive manufacturing process PBF-LB/M.
- Density, microstructure and hardness depth profiles are of good quality and comparable to conventional smelting-metallurgy gears.
- The lightweight hub designs have no negative effect on the load carrying capacity and showed a high potential for further mass reduction (m_{red}>45%).

The outlook provides the following topics:

- Optimized lightweight gear hub designs should aim for further mass reductions (e.g. $m_{red} \geq 60\%$).
- Further lightweight potential can be exploited by additively manufacturing gear and shaft together in one PBF-LB/M part.

References

1. Thompson, M., Moroni, G., Vaneker, T., Fadel, G., Campbell, R. I., Gibson, I., Bernard, A., Schulz, J., Ahuja, B., and Martina, F., Design for Additive Manufacturing: Trends, oppurtunities, considerations, and constraints, CIRP Annals - Manufacturing Technology, 65, pp. 737 – 760, 2016.
2. Schmitt, M., Kamps, T., Siglmüller, F., Winkler, K. J., Schlick, G., Seidel, C., Tobie, T., Stahl, K., and Reinhart, G., Laser-based powder bed fusion of 16MnCr5 and resulting material properties, Additive Manufacturing, 35, 2020. DOI: https://doi.org/10.1016/j.addma.2020.101372
3. Siglmüller, F., Kupfer, S., Kamps, T., Schmitt, M., Lohner, T., Reinhart, K., and Stahl, K., Efficiency of additive manufactured gears with conformal cooling, 21st TAE International Colloquium Tribology, 2018.
4. Schmitt, M., Jansen, D., Bihlmeir, A., Winkler, K. J., Anstätt, C., Schlick, G., Tobie, T., Stahl, K., and Reinhart, G., Rahmen und Strategien für den Leichtbau von additiv gefertigten Zahnrädern für die Automobilindustrie, Rapid.Tech + FabCon 3.D International Hub for Additive Manufacturing: Exhibition + Conference + Networking, pp. 89 – 102, 2019. DOI: https://doi.org/10.3139/9783446462441.007
5. VDI-Fachbereich Produktionstechnik und Fertigungsverfahren, Additive Fertigungsverfahren; Strahlschmelzen metallischer Bauteile; Qualifizierung, Qualitätssicherung und Nachbearbeitung, VDI (3405 Blatt 2). VDI-Richtlinie, 2013.
6. International Organization for Standardization (ISO), Calculation of load capacity of spur and helical gears (including all current standards, technical specifications and technical reports), ISO (6336). International Standard. Beuth Verlag GmbH, Berlin, 2016, 2019.
7. Bruzek, B. and Leidich, E., Eds., 2007, Dünnwandige verzahnte Naben mit Passfedernut. VDI-Berichte
8. Fuchs, D., Schurer, S., Tobie, T., and Stahl, K., On the determination of the bending fatigue strength in and above the very high cycle fatigue regime of shot-peened gears, Forschung im Ingenieurwesen, 2021. DOI: https://doi.org/10.1007/s10010-021-00499-2
9. Kratzer, D., König, J., Tobie, T., and Stahl, K., Eds., 2020, Effects of different shot peening treatments in combination with a superfinishing process on the surface durability of case-hardened gears
10. Schmitt, M., Winkler, K. J., Schlick, G., Tobie, T., Reinhart, G., and Stahl, K., Abschlussbericht: Integraler Leichtbau bei Zahnrädern mittels Laserstrahlschmelzen. Deutsche Forschungsgemeinschaft (DFG), Augsburg und München, 2021.
11. Leonhardt, C., Benkert, T., Meißner, R., Nadolski, D., Otto, M., and Stahl, K., Intelligenter Leichtbau durch Mehrkomponentenverfahren, Frankfurt, 2020.
12. Niemann, G., Winter, H., and Höhn, B.-R., 2003, Maschinenelemente - Band 2: Getriebe allgemein, Zahnradgetriebe - Grundlagen, Stirnradgetriebe; *Machine Elements - Volume 2: General Transmissions, Gearboxes - Fundamentals, Spur Gears*. Springer Berlin Heidelberg

13. Koller, P., Tobie, T., and Höhn, B.-R., Optimierung Flankentragfähigkeit – Steigerung der Zahnflankentragfähigkeit durch Kombination von Strahlbehandlung und Finishingprozess. FVA-Heft 957. Forschungsvereinigung Antriebstechnik e.V., Frankfurt/Main, 2010.
14. Felbermair, M., Tobie, T., and Stahl, K., Einfluss der Graufleckigkeit auf die Grübchentragfähigkeit einsatzgehärteter Zahnräder im Zeit- und Dauerfestigkeitsbereich. FVA-Heft 1087. Forschungsvereinigung Antriebstechnik e.V., Frankfurt/Main, 2014.

3D Material Model for Additive Manufactured Metallic Parts

Emre Ertürk[1(✉)], Jens Bold[2], Philipp Höfer[1], Christoph Stark[3], and Wolfgang Höhn[4]

[1] Institute of Lightweight Engineering, Universität der Bundeswehr München, 85577 Neubiberg, Germany
{emre.ertuerk,philipp.hoefer}@unibw.de
[2] Boeing Deutschland GmbH, 80805 Munich, Germany
jens.bold@boeing.com
[3] Modell- Und Formenbau, Blasius GERG GmbH, 83104 Hohenthann, Germany
Christoph.Stark@gerg.de
[4] GMA-Werkstoffprüfung GmbH, 86167 Augsburg, Germany
w.hoehn@gma-group.com

Abstract. Additive manufactured parts have a high potential in weight saving for aerospace applications. In the Bavarian-funded project BayLu25 – BLANCA, the Universität der Bundeswehr München, the test house GMA, the small and medium enterprise GERG and the aircraft manufacturer Boeing are working together to investigate and develop additive manufactured load introductions into sandwich parts. Certification is taken as a basis for the developments and therefore, 3D simulation methods are used to verify and validate the structure. Current metal parts are commonly certified using the von-Mises yield criterion. The advantages in higher compression and shear allowable as well as the orthotropic behavior of additive manufactured metals are not considered. To further decrease weight and therefore save CO_2, the 3D material behavior of Ti-6Al-4 V is investigated in the current project. The material characterization was done by means of a test program, in which the 3D material properties for tension, compression and shear loading were determined. Furthermore, the build orientation influence on the strength and stiffness values of the specimens was evaluated. For modeling the material behavior, the Cuntze-Bold model was adapted for metallic parts.

Keywords: Additive Manufacturing (AM) · Metal · Testing · Material Modeling

1 Introduction

Due to the high design freedom AM offers, conventionally manufactured parts are progressively replaced by additive manufactured structures. In the Bavarian-funded project BayLu25-BLANCA, the Universität der Bundeswehr München, the test house

J. Rieser et al. (Eds.): *Proceedings of the Munich Symposium on Lightweight Design 2021*, Proceedings, pp. 175–188, 2023. https://doi.org/10.1007/978-3-662-65216-9_16

GMA, the small and medium enterprise GERG and the aircraft manufacturer Boeing are collaborating to optimize bonded load introduction into sandwich components using additive manufactured load introduction structures. One way to increase the lightweight potential is to further develop the AM-Process. Another possibility is to improve the calculation methods used for strength analysis during certification. Thus, accurate modeling of the material behavior, especially the failure criterion for strength evaluation, is important for the resulting weight. Therefore, advanced simulation methods are developed in this project, which shall contribute to simulation-assisted certification.

In aerospace, certification is done according to regulations issued by e.g., the European Union Aviation Safety Agency (EASA) [1]. These regulations state that reduced material allowables must be used for strength calculation. For this purpose, a distinction is made between A- and B-values, as indicated in Table 1.

Table 1. Definition of A- and B-Values [2, 3]

	Load path design	Probability basis p	Material allowable
A-value	single	99%	$\overline{x} - k_p \cdot \sigma$
B-value	multi	90%	$\overline{x} - k_p \cdot \sigma$

In the research project BLANCA, a single load path design is developed. Therefore, the A-value must be used. This ensures, that at least 99% of the population of values is equal to or exceeds the A-basis with a confidence of 95%. A-value calculation is done by reducing the arithmetic mean material value obtained from testing, with a reduction factor that consists of a reduction parameter k_p and the standard deviation σ. In addition to the experimental inaccuracies, the manufacturing accuracy of the specimens can also influence σ, meaning that it can have a significant impact on the resulting material allowables used for certification. Particularly in AM, it is therefore important to determine the material values and their scattering for different build orientations. In this paper, this was done for a laser powder bed fusion (LPBF) process. The influence of the manufacturing accuracy on the material allowables is discussed. Furthermore, a newly developed material and failure model is presented, which was evaluated using the generated material data. This approach considers the build orientation and load direction dependent material allowables for strength calculation. Finally, an outlook is given on further necessary steps to verify the failure criterion.

2 Material Characterization

In this study the titanium alloy Ti-6Al-4V was investigated. The material characterization was done by means of a test program, in which the material behavior was determined for tension, compression and shear loading. ASTM standards were used as a basis for testing. The given specimen geometries were applied for tensile and shear tests, while for the compression tests a modified geometry was defined. In addition to the load direction (tension, compression, shear), the dependence of the material values on the specimen's build orientation was also considered. Figure 1 shows the investigated build orientations. Specimens built in vertical direction were labeled with T, while for specimens aligned in longitudinal and width direction the labels L and W were defined.

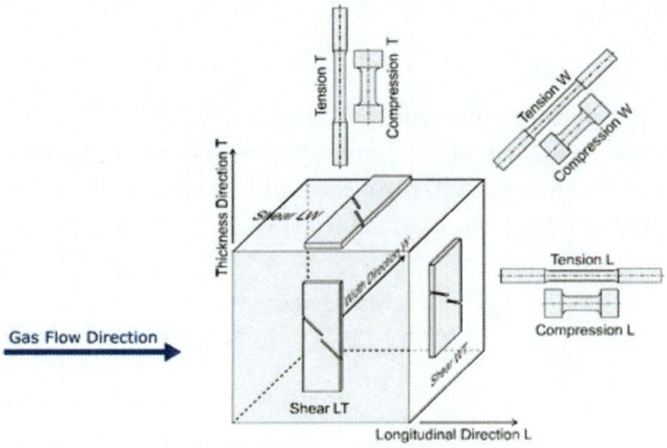

Fig. 1. Definition of the build orientations

Eight specimens were tested for each load direction and build orientation, resulting in a total of 72 specimens used for the characterization of one material. Manufacturing was done by the project member GERG, using a Concept Laser M2 and minimum postprocessing steps. After unpacking the specimens, a stress-relief heat treatment was applied. An extra machining step was only necessary for the compression specimens, in which the end faces were machined to ensure parallel level surfaces for load introduction. Furthermore, it should be noted that due to the limited build plate size, the specimens had to be printed in two batches. The build job of one batch is shown in Fig. 2.

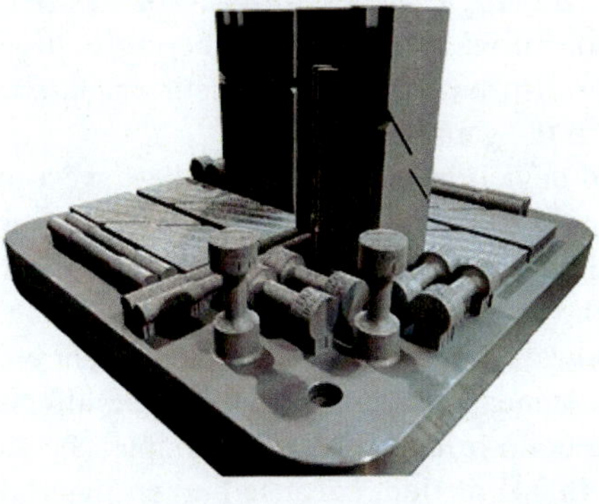

Fig. 2. Build job of Ti-6Al-4 V specimens

In the following subchapters, the experimental results from tensile, compression and shear testing are presented and the influence of the build orientation on the material strength is shown. Testing was performed by the project member GMA on a universal testing machine (Zwick Z330). The environmental conditions were set to room temperature (23 °C ± 2 °C) and 50% ± 5% humidity. To support these results, additional testing was conducted at the Universität der Bundeswehr München, using a digital image correlation system for compression and shear tests.

2.1 Tensile Tests

Tensile testing was performed using the standard ASTM A370. A crosshead speed of 10 MPa/min was defined and an extensometer was used for strain measurement. Evaluating the stress–strain-curves for each build orientation L, W, and T, resulted in the following ultimate tensile strength values (Fig. 3).

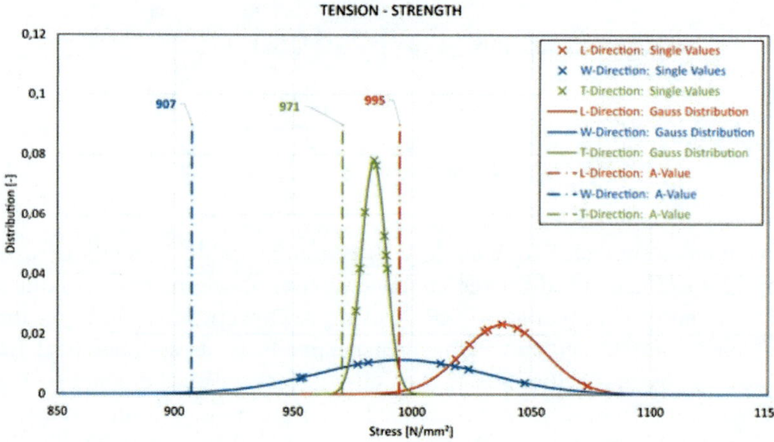

Fig. 3. Tensile strength values for the build orientations L, W, T. Single- and A-values are shown

In this graph, the eight single values for each build orientation and the resulting A-values, used for certification, are shown. Furthermore, it was assumed that the single values are Gaussian distributed. The arithmetic mean strength value is indicated here, by the maximum of the gauss distribution.

The lowest standard deviation, and therefore the lowest material value scattering, was found for specimens oriented in T-direction. It amounts to 0.52%, while the largest standard deviation is in W-direction with 3.4%. The mean strength values are on a similar level, while the strength in L-direction tends to be higher. However, comparing the A-values shows a more pronounced dependency on the build orientation. This is due to different standard deviations, which are affected by varying manufacturing accuracy in each build orientation. For example, the strength in W-direction is about 6.5% lower than in T-direction, because higher material value scattering causes a larger reduction of the mean value. The same evaluation was conducted for the specimen's elastic stiffness values. A similar influence of the build orientation could be observed here. To consider this influence on A-values, a 3D failure criterion should be used for strength calculation.

2.2 Compression Tests

Compression tests were conducted based on ASTM E9. A modified specimen shape was defined as shown in Fig. 4.

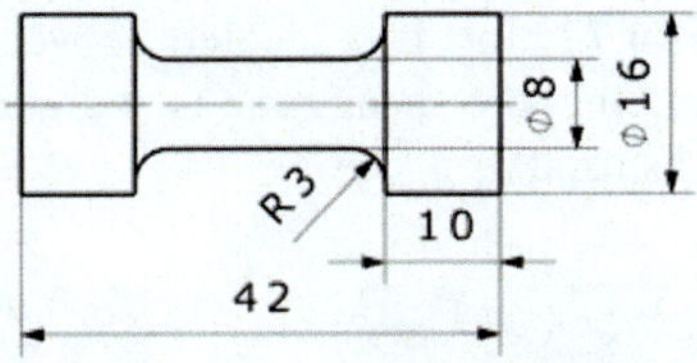

Fig. 4. Compression specimen (Dimensions in mm)

The diameter at the ends was chosen larger, to ensure material strength failure. Moreover, it is advantageous for reducing load introduction effects in the measurement area [4]. The testing speed was set to 2 mm/min and for measuring the longitudinal strains a strain gauge was applied to the center area of the specimen. The evaluated compression strength values are depicted in the following figure (Fig. 5).

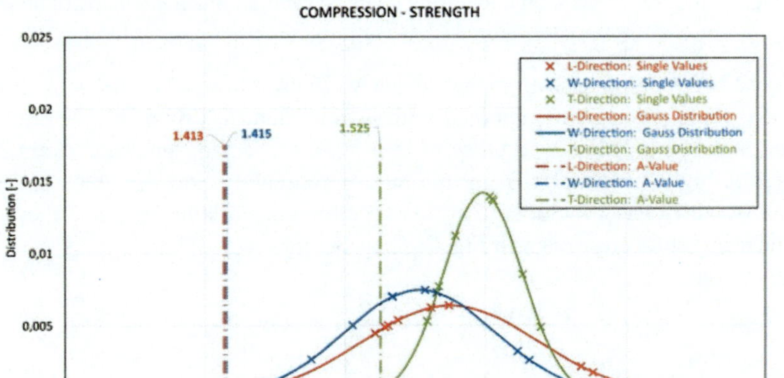

Fig. 5 Compression strength values for the build orientations L, W, T. Single- and A-values are shown

The mean strength values are close to each other, while the A-values in L- and W-direction are about 8% lower than in T-direction. This is due to higher standard deviation in these two directions. It can be noted that the lowest material value scattering was again found for specimens oriented in T-direction. Furthermore, comparing these results with the tensile strength values showed that the compression strength is at least 42% higher. This indicates that the load direction should be considered in a material- and failure model.

2.3 Shear Tests

For shear testing, the standard ASTM B831 was used. The given specimen has a small shear zone with a length of 4.72 mm. It is located between two notches as shown in Fig. 6. To investigate how distinct the shear zone is, a digital image correlation system (Q400D, Limess GmbH) was used.

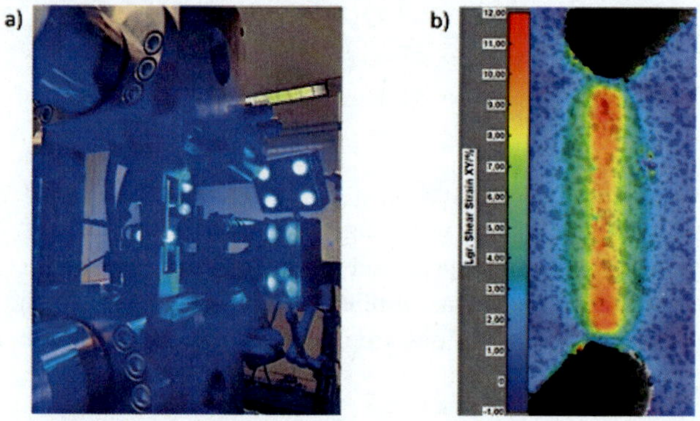

Fig. 6. **a**) Shear zone evaluation using a DIC-system, **b**) Shear strain distribution

Two 12 MPix cameras with macro lenses were used to resolve the shear zone. The required speckle pattern was obtained with a spray gun. In Fig. 6 (b), the shear strain distribution is shown for a load close to failure. As expected, the shear strain close to the edges is higher, but apart from this an approximately homogeneous shear strain field is formed. It was concluded that this specimen is suitable for shear characterization. The test results are shown in Fig. 7.

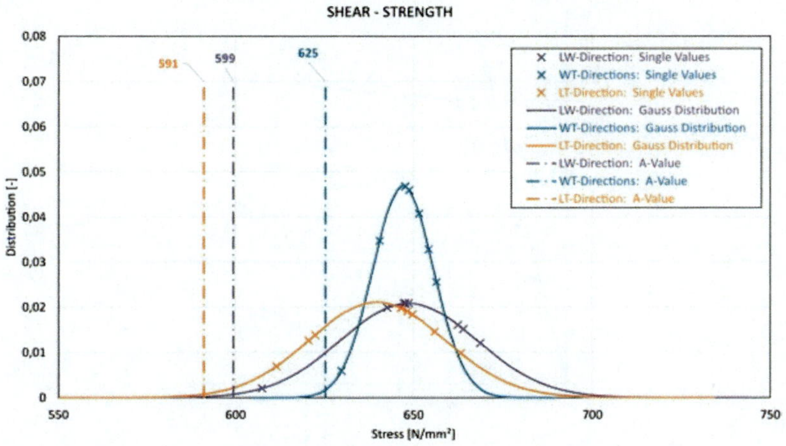

Fig. 7. Shear strength values for the build orientations L, W, T. Single- and A-values are shown

The smallest standard deviation was again found for specimens oriented in T-direction. It can be concluded that the highest accuracy for the material strength and stiffness properties can be achieved for specimens oriented in vertical direction. When comparing the mean shear strength values, no significant difference was found. Therefore, an isotropic material behavior could be assumed. However, for the A-values a dependency on the build orientation is present. This indicates that for structural design a 3D material- and failure model could be useful. Furthermore, it was noted that the measured shear strength value is about 10% higher than the one which can be calculated using the von-Mises yield-criterion [5]. The lightweight potential in structural design could therefore be increased using a failure criterion that considers load direction dependent material allowables.

3 Non-linear Material Model and Failure Criterion

This chapter will describe the new BOLD non-linear material model and CUNTZE failure criterion for additive manufactured materials. Combined it will be named CUNTZE-BOLD AM.

3.1 Non-linear Material

The material behavior before yield follows Hooke's law:

$$\sigma_{BY}(\varepsilon) = E^{in} \cdot \varepsilon \tag{1}$$

The past yield behavior (Fig. 8) can be approximated in a first step as linear and described with the yield stress $R_{p0.2}$, the correlated strain $\varepsilon_{p0.2}$, the ultimate strength R_m and the corresponding ultimate strain ε_m as shown in Eq. (2), where $\varepsilon_{p0.2} = \frac{R_{p0.2}}{E^{in}} + \varepsilon_y$.

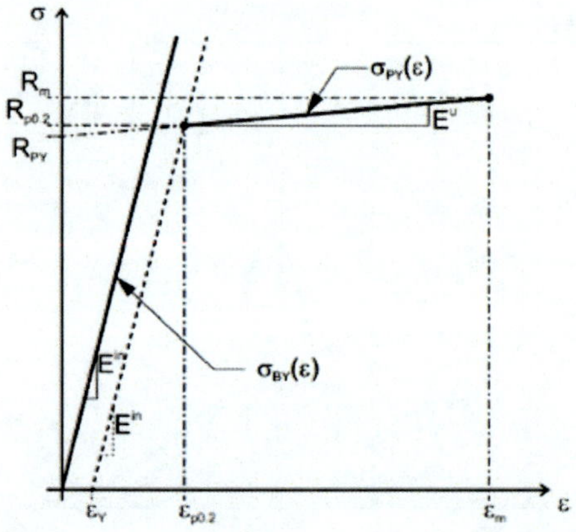

Fig. 8. Linear material behavior past yield point

$$\sigma_{PY}(\varepsilon) = R_m + E_u \cdot (\varepsilon - \varepsilon_m) = R_m + \frac{R_m - R_{p0.2}}{\varepsilon_m - \varepsilon_{p0.2}} \cdot (\varepsilon - \varepsilon_m) \tag{2}$$

The strain at the intersection point ε_i (Fig. 9) can be determined by setting the equations for the description before yield $\sigma_{BY}(\varepsilon)$ and linear past yield $\sigma_{PY}(\varepsilon)$ equal (Eq. (3)) and solving for the strain (Eq. (4)).

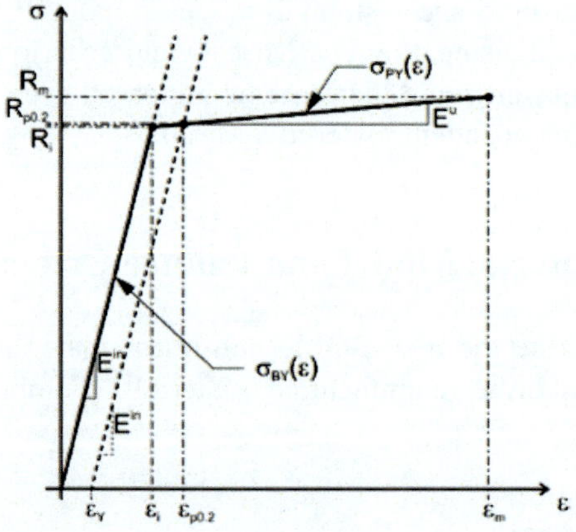

Fig. 9. Intersection between linear elastic and linear past yield behavior

$$\sigma_{BY}(\varepsilon) = \sigma_{PY}(\varepsilon) \tag{3}$$

$$\varepsilon_i = \frac{R_m - E_u \cdot \varepsilon_m}{E^{in} - E_u} \tag{4}$$

Compared to the classically used Ramberg–Osgood material model [6] where the strain is described by the stress, the new BOLD non-linear material model uses the strain to calculate the stress. This method is more convenient for engineers and can be easily programmed. No if–then equation is used, but a step function to transfer from the behavior before yield to the past yield behavior (Eq. (5)). In Fig. 10, the resulting function $\sigma(\varepsilon)$ is compared with the test results for tension.

$$\sigma(\varepsilon) = E^{in} \cdot \varepsilon \cdot \left(\frac{\sigma_{PY}}{\sigma_{BY}}\right)^{g(\varepsilon)} \tag{5}$$

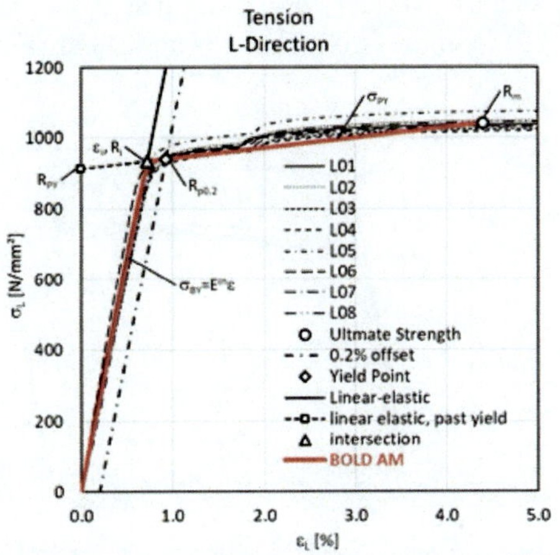

Fig. 10. Typical stress–strain diagram in tension L-direction

The step function $g(\varepsilon)$ is based on an analytical function – the hyperbolic tangent – which runs from -1 for minus infinite to+1 for plus infinite. It is defined by the e-function and is used here in the following form:

$$g(\varepsilon) = \frac{1}{e^{2(C+D\varepsilon)} + 1}. \tag{6}$$

The two factors C and D can be calculated using Eq. (7) and (8). The values 0.99 and 0.01 are the accuracies for the lower (0.01) and the upper (0.99) stress. The function will switch between the lower strain value of $2 \cdot \varepsilon_i - \varepsilon_{p0.2}$ and $\varepsilon_{p0.2}$ with a width of $2 \cdot \left(\varepsilon_{p0.2} - \varepsilon_i\right)$ as shown in Fig. 11.

$$D = \frac{ln\left(\frac{1}{0.99} - 1\right) - ln\left(\frac{1}{0.01} - 1\right)}{2 \cdot \left(\varepsilon_{p0.2} - \varepsilon_i\right)} \tag{7}$$

$$C = ln\left(\frac{1}{0.99} - 1\right) - D \cdot \varepsilon_{p0.2} \tag{8}$$

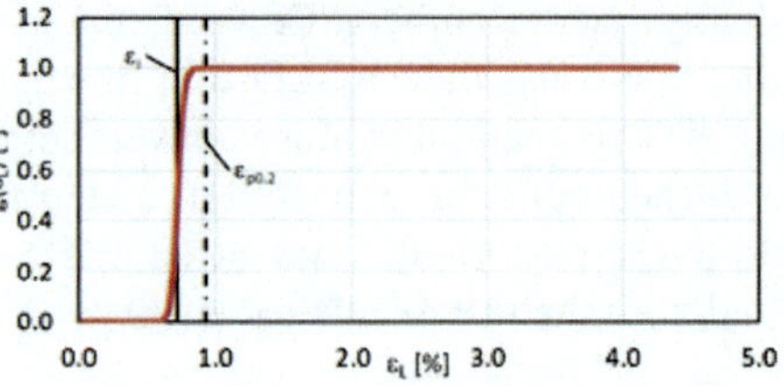

Fig. 11. Transition function

The BOLD non-linear material model can be applied to all build orientations (L, W and T) in tension (Fig. 12), compression (Fig. 13) and shear (Fig. 14) loading.

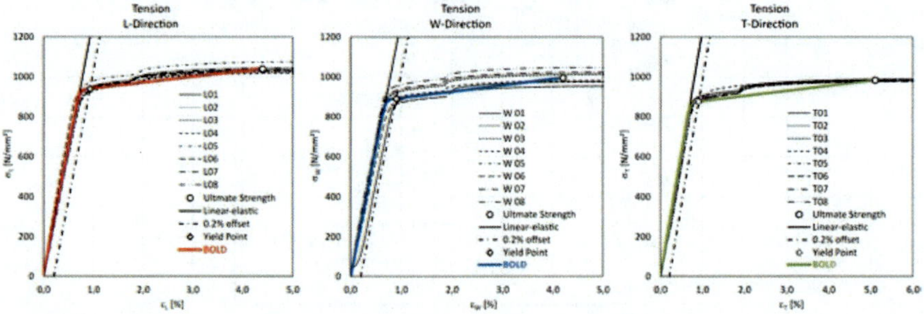

Fig. 12. BOLD non-linear material model applied to tension in L-, W- and T-direction

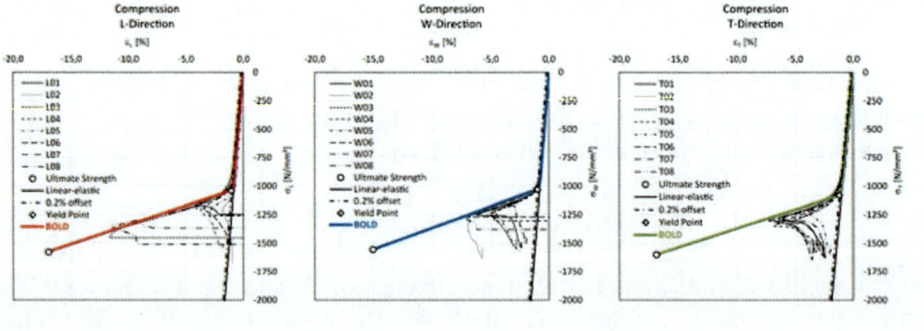

Fig. 13. BOLD non-linear material model applied to compression in L-, W- and T-direction

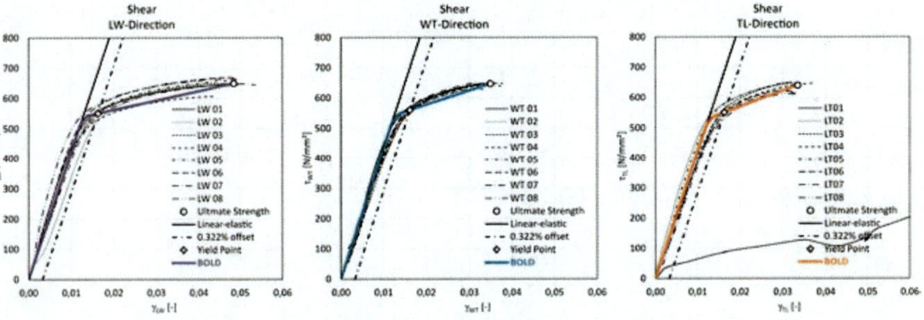

Fig. 14. BOLD non-linear material model applied to shear in LW-, WT- and TL-direction

3.2 Failure Criterion

The classical failure criterion being used for metal materials is the von-Mises yield criterion as shown in Eq. (9). Only the tension allowable is used to determine the von-Mises stress. No different build orientations (L-, W- or T-direction) or loading directions (tension, compression and shear) are considered.

$$\sigma_{VM} = \sigma_L^2 + \sigma_W^2 + \sigma_T^2 - \sigma_L\sigma_W - \sigma_W\sigma_T - \sigma_T\sigma_L + 3\tau_{LW}^2 + 3\tau_{WT}^2 + 3\tau_{TL}^2 \leq R_m \quad (9)$$

For orthotropic materials, in general composites, Eq. (10) can be used according to the Tsai-Wu criterion [7], which is considering the different build and load direction. Unfortunately, it is only available for 2D stress distribution and does not include the third direction.

$$\frac{\sigma_L^2}{R_L^t R_L^c} + \frac{\sigma_W^2}{R_W^t R_W^c} - \left(\frac{1}{R_L^t} - \frac{1}{R_L^c}\right)\sigma_L - \left(\frac{1}{R_W^t} - \frac{1}{R_W^c}\right)\sigma_W + F_{12}^* \frac{\sigma_L\sigma_W}{\sqrt{R_L^t R_L^c R_W^t R_W^c}} + \frac{\tau_{LW}^2}{R_{LW}^2} \leq 1 \quad (10)$$

To overcome these challenges, a new failure criterion, called CUNTZE-BOLD, was developed based on the CUNTZE failure criterion for woven materials [8, 9]. The build orientations will be considered as well as different loading directions. Normal loadings (Eq. (11–13)) were named N_i with the index i describing the direction ($i = L, W, T$). The shear loading (Eq. (14–16)) was named S_{ij} with the index ij for the shear plane ($ij = LW, WT, TL$). For all types, the value is calculated by the ratio between the calculated stress, σ_i for normal and τ_{ij} for shear stress, and the corresponding allowable. Although these values look very similar to efforts in the single direction, they should not be understood as these, because efforts should be within a value of 0 to 1. The values in normal direction N_i can vary from $-2/\sqrt{3}$ to 0 in compression and 0 to $2/\sqrt{3}$ in tension. Finally, the single loadings are combined in the failure criterion as shown in Eq. (17).

$$N_L = \begin{cases} \frac{\sigma_L}{R_L^t} for\ \sigma_L \geq 0 \\ \frac{\sigma_L}{R_L^c} for\ \sigma_L < 0 \end{cases} \quad (11)$$

$$N_W = \begin{cases} \frac{\sigma_W}{R_W^t} for\ \sigma_W \geq 0 \\ \frac{\sigma_W}{R_W^c} for\ \sigma_W < 0 \end{cases} \tag{12}$$

$$N_T = \begin{cases} \frac{\sigma_T}{R_T^t} for\ \sigma_T \geq 0 \\ \frac{\sigma_T}{R_T^c} for\ \sigma_T < 0 \end{cases} \tag{13}$$

$$S_{LW} = \frac{|\tau_{LW}|}{R_{LW}} \tag{14}$$

$$S_{WT} = \frac{|\tau_{WT}|}{R_{WT}} \tag{15}$$

$$S_{TL} = \frac{|\tau_{TL}|}{R_{TL}} \tag{16}$$

$$N_L^2 + N_W^2 + N_T^2 - N_L N_W - N_W N_T - N_T N_L + S_{LW}^2 + S_{WT}^2 + S_{TL}^2 \leq 1 \tag{17}$$

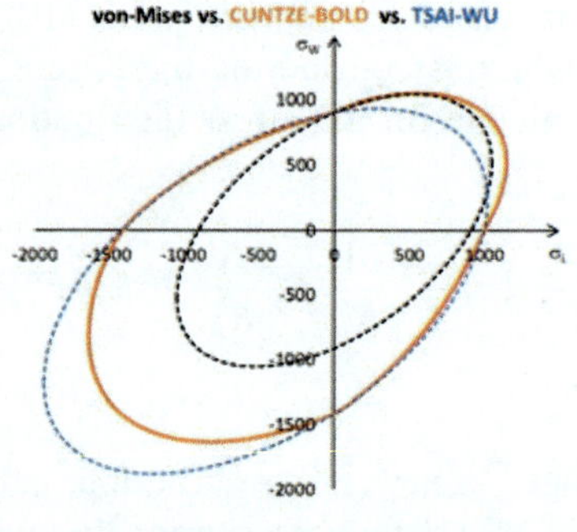

Fig. 15. Comparison of normal-normal failure envelope

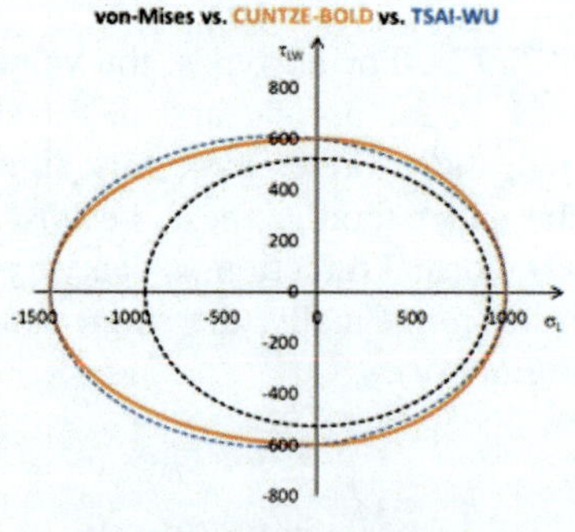

Fig. 16. Comparison of normal-shear failure envelope

Figure 15 shows the three failure envelopes for von-Mises, Tsai-Wu and CUNTZE-BOLD for the normal-normal stress in σ_L and σ_W. This comparison indicates that the von-Mises yield criterion is conservative because it only takes the tension allowable as reference. The Tsai-Wu failure criterion is even more conservative in the first sector (tension-tension loading) while in the third sector (compression-compression loading) it is optimistic. The failure envelope of the CUNTZE-BOLD failure criterion is in the third sector between both criteria, while in the first sector similarities to the von-Mises yield criterion are present. To assess each failure criterion, biaxial testing must be conducted.

This comparison is also shown for a normal-shear stress combination (σ_L-τ_{LW}) in Fig. 16. The von-Mises yield criterion is again the conservative one because here it also only takes the tension allowable as reference. The Tsai-Wu failure and CUNTZE-BOLD failure criterion are very similar in this case. Only two failure envelopes were shown here. It should be noted that in total 15 envelopes must be considered: three for normal-normal stress combination, nine for normal-shear combination and three for shear combination.

4 Conclusion and Outlook

A material characterization was conducted for additive manufactured Ti-6Al-4V, by means of a test program. The influence of the build orientation on the material strength and stiffness was investigated for tension, compression and shear loading. It could be shown that the A-values are dependent on the build orientation, because of different reproducibility of the specimens in L, W, and T-direction. The lowest material value scattering was observed for specimens oriented in vertical direction, regardless of the load direction. Furthermore, test evaluation showed that a failure criterion, which considers material allowables for each load direction could be favorable for the investigated material. For modeling the material behavior, the Cuntze-Bold-model was adapted for additive manufactured metallic parts. Comparing the failure envelopes indicated that the developed failure criterion is less conservative than the von-Mises yield-criterion and it also differs from the Tsai-Wu model. In subsequent studies, biaxial testing will be conducted to qualify the failure criterion. In addition, further metals such as AlSi10Mg will be tested for model qualification.

Acknowledgements. This study is part of the research project BLANCA, which is funded by the Bavarian Ministry of Economic affairs, Regional development and Energy (StMWi) within the program BayLu-25.

Sponsored by

Bavarian Ministry of Economic Affairs, Regional Development and Energy

References

1. European Union Aviation Safety Agency: Certification Specifications and Acceptable Means of Compliance for large Aeroplanes CS-25. Amendment 26 (2020).
2. European Union Aviation Safety Agency: Acceptable Means of Compliance AMC 20–29, Appendix 2 (2020).
3. United States Department of Defense: Metallic Materials and Elements for Aerospace Vehicle Structures, MIL-HDBK-5C. Vol 1, Revision C (1976).
4. Riebensahm, P., Träger, L.: Werkstoffprüfung (Metalle). Springer, Heidelberg (1928).
5. Mises, R. v.: Mechanik der festen Körper im plastisch-deformablen Zustand. Nachrichten der königlichen Gesellschaft der Wissenschaft in Göttingen, Mathematisch-Physikalische Klasse, 582 (1913).
6. Ramberg, W., Osgood, W.R.: Description of stress-strain curves by three parameters. NACA, Washington, D.C. (1943).
7. Tsai, S. W., Wu, E. M.: A General Theory of Strength for Anisotropic Materials. Journal of Composite Materials, 5, 58-80 (1971).
8. Cuntze, R.: Comparison between experimental and theoretical results using Cuntze's failure mode concept model for composites under triaxial loadings. Journal of Composite Materials, SAGE Publications (2012).
9. Bold, J.: Vergleich des Impaktverhaltens von monolithischer und hybrider CFK-Platte unter Verwendung eines neuen Werkstoffmodells. PhD Thesis, Deutsches Zentrum für Luft- und Raumfahrt e.V., Köln (2019).

Structural Optimization in Lightweight Design for SLM Meets Additive Serial Production and Efficient Post-Machining

Rinje Brandis[1](✉), Martin Blanke[1], and Jan Rams[2]

[1] DMG MORI Academy GmbH, Gildemeisterstraße 60, 33689 Bielefeld, Germany
{rinje.brandis,additive.intelligence, martin.blanke}@dmgmori.com

[2] Technische Hochschule OWL, Campusallee 12, 32657 Lemgo, Germany
Jan.rams@dmgmori.com

Abstract. Additive manufacturing (AM) is not a technology that can be viewed solitary, but in return it is increasingly collaborating with upstream such as CAE software and downstream technologies such as subtractive manufacturing. Even more important, this technology must collaborate in order to leverage all the advantages that this technology offers.

Using the example of the robot head of the DMG MORI "Robo2Go", a flexible modular robot automation unit and a serial product of the portfolio of DMG MORI, a holistic redesign was carried out that considers the entire process chain. The goal was to optimize the system accuracy, reduce the assembly time and achieve a lightweight design of the assembly. In the context of the redesign of the robot head, methods and strategies for modeling and analyzing lightweight structures were developed and demonstrated by using innovative simulation software and through close interlinking and early consideration of mechanical post-processing with CNC machine tools.

This project illustrates the potential of a holistic product development not only with a focus purely on AM, but also with a focus on the complete production process up to the final lightweight product.

In this redesign project, the combination of AM and milling resulted in an accuracy that is 16 times higher than the original milling design. Furthermore, the assembly time is significantly reduced, the assembly weight is minimized by up to 64%,the new robot head only has 20 sealing points—45 percent less than before, the robot head originally consisted of 135 individual parts and the AM component reduces the number to 56 parts, while the production costs have remained almost identical.

Keywords: Structural Optimization · Lightweight Design · SLM · Additive serial production · Post-machining

J. Rieser et al. (Eds.): *Proceedings of the Munich Symposium on Lightweight Design 2021*, Proceedings, pp. 189–197, 2023. https://doi.org/10.1007/978-3-662-65216-9_17

1 Introduction

The industrial product development and manufacturing process is transforming into an agile, rapid, and economical process. Determining factors in this transformation are market and customer demands such as time reduction and cost minimization. Factors that also affect lightweight structures, an area that has received a technology push due to the design and manufacturing possibilities of Additive Manufacturing. [1]

Additive Manufacturing and in particular the SLM-process (Selective Laser Melting) is not a technology that can be viewed on its own, but in return it is increasingly collaborating with upstream and downstream technologies. Even more important, this technology must collaborate in order to leverage all the advantages that this technology offers.

A powerful and efficient value chain consist of design, manufacturing and post-machining steps. At first, the design phase in which CAD and similar software solutions are used to create function integrated and topology optimized parts resulting in complex technical designs as preparation for the manufacturing process. Secondly, the actual production is carried out by one of the various AM technologies available, in the presented case the SLM-process. Subsequently, a subordinate process chain usually follows, the so-called post-processing. Hereby the quality of the additively created blank is improved by means of suitable processes to enhance the surface, properties or applying technical connections by CNC machining.

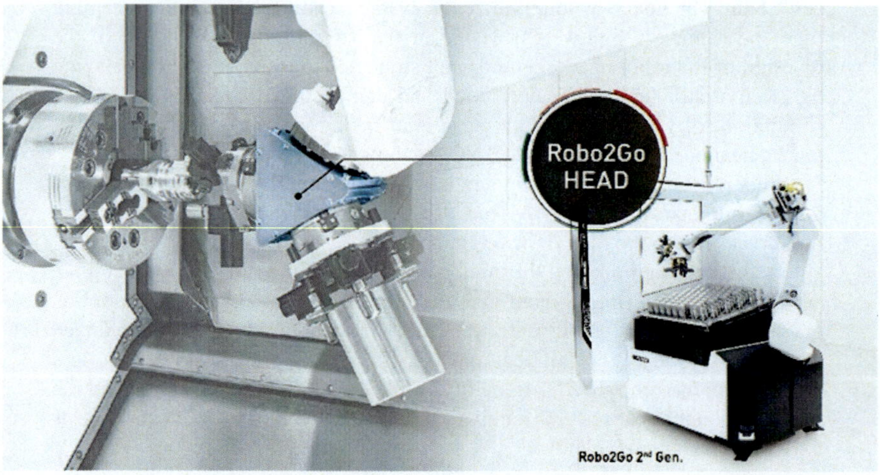

Fig. 1 Robo2Go—work environment

This paper discusses the transformation by using the project of the "Robo2Go", in which a holistic redesign was carried out that considers the entire value chain. The innovative Robo2Go (Fig. 1), a serial product of the portfolio of DMG MORI, offers a flexible automation layout with free access to the machine as well as an intelligent safety concept for the collaboration of man and machine. High flexibility is achieved by simple handling and the possibility of free placement for several different lathes in short time. The head is a structural component as link between the robot and the gripping device that eventually carries the workpieces.

2 Challenges

Each stage of the development process is characterized by its particular challenges and restrictions. For the design, existing connections and measurements must be combined with the aims of the optimization process: functional integration and weight reduction [2]. Therefore, the design task is rather complex and additionally requires the usage of coexistent software solutions. Inconsistent data interfaces impede continuous processes. In literature the term "DfAM—Design for Additive Manufacturing" was established to express and dedicate emphasis to this field of research [3]. In parallel cost and duration of development ought to be reduced to reach competitive products in shorter time. Lightweight design is an interdisciplinary engineering task between structural design, material science, simulation software and production technology.

The additive production followed directly after the design phase, which underlies its own complexity and challenge. The component, the machine and parameters must be aligned to produce usable parts. Especially topology optimized parts often do not meet manufacturability and require support material. Heat distortion through the laser welding process possibly causes dimensional deviations and bad surface quality. To counter this, the machine settings must be selected correctly and adapted to the material and component.

Even if design and production phase create nearnet shape results the current state of additive manufactured parts predominantly and inevitably require downstream post-processing to meet dimensional and geometric tolerances set by the customer or the application. [4] This fact is often neglected and not well discussed in literature but must be considered already in the design phase of the component. Else, the nature of lightweight, the reduction of unnecessary material, causes major problems for the machinability, e.g. for CNC milling, because it results in difficult and unstable clamping and processing conditions.

In short, the holistic approach to DfAM or redesign projects is only possible through interconnecting up- and downstream development stages and gathering all necessary production information in the design phase to reduce errors afterwards.

3 Process Overview

In the context of the redesign process of the robot head, methods and strategies for modeling and analyzing lightweight structures were developed and through close interlinking and early consideration of mechanical post-machining with CNC machine tools. A powerful and efficient process resulted mature of design, manufacturing, and post-machining phase.

3.1 Design

The goal was to optimize the quality, reduce the assembly time and achieve a lightweight design of the assembly. In the context of additive redesign, a new and holistic mindset is necessary to be successful. Structural lightweight design is the ideal strategy, as its goal is maximum material utilization through an optimized and ideal shape. The goal can be achieved, among other things, by applying the design principles for additive manufacturing and lightweight and stress-oriented design. [5]

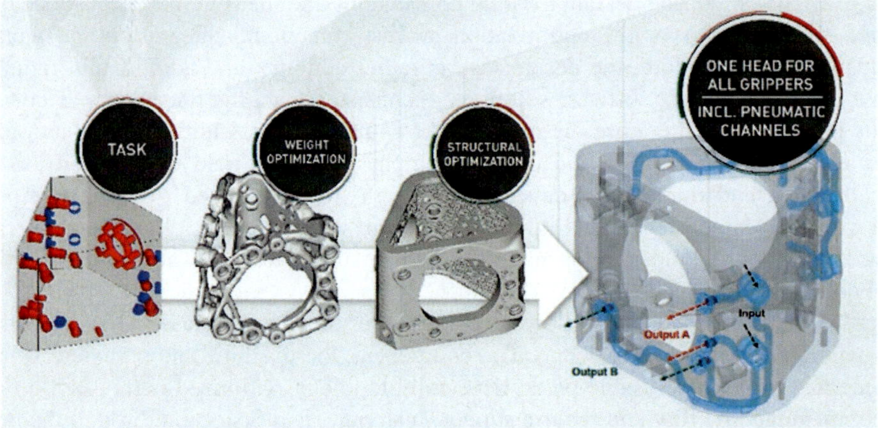

Fig. 2 Robo2Go—Redesign with functional integration and topology optimization

Already during the first additive quick check, a high additive potential of the assembly could be identified; in particular, the functional integration, the direct introduction of force, material savings in areas of low loads, and full utilization of construction should be mentioned (Fig. 2). In the design process, 79 conventional individual parts were transformed into a highly complex integral design for additive manufacturing. A new design methodology to realize a variable shelling and latticing was developed to reduce the weight of the integral component. The field-driven design capabilities of the software "nTopology" is the core of the new methodology. In the following, the methodology will be described step-by-step. First, design automation based on Color-Coding was developed and implemented into the design process. The benefits of automation is a robust and reusable workflow for using the color property to select relevant surfaces and function elements.

In a traditional CAD system, all surfaces of the original design were color-coded for each subfunction and -systems. Each color represents a different subsystem: blue for the pneumatic inlets and outlets, yellow for the pneumatic channels, red for the interfaces with the robot, purple for mounting points, green for contact surfaces, and white for the external surfaces. If the geometry changes in the future, only the new CAD file must be imported, and the design and analysis process would rerun automatically. Of course, the workflow can easily be adapted and reused for variant designs.

Secondly, each colored surface has an individual thickness assigned before combining them into a single body. The thickness of the outside surface was driven by topology optimization to maximize the stiffness with minimal use of material. Here the design automation based on Color-Coding was applied as well. Using the color-coded surfaces as input the design space and forces for the topology optimization were defined automatically. The topology optimization result was then utilized as input to create an optimized thickness of the outside surface: thicker where more material was required and thinner where it was not. The thickness of areas where the result contacted the outer shell should be thicker than the non-load-bearing areas of the frame. This process allowed to use the structure benefits of topology optimization without changing the outside surface. Once the exterior thickness was optimized, function elements like the interface between robot and robot head or the pneumatic system also needed to be lightweight and integrated. Here the design automation based on Color-Coding is used as well. Using the color-coded surfaces as input the thickness of the function elements has been applied automatically. Once the separate exterior and the function elements were generated, all could be combined into a single body using a simple boolean union. The Boolean union can join many single bodies into one, independent of geometric complexities. Optionally, a fillet can be added between the bodies.

The single body had a lot of areas with large overhangs so that the body could not be manufactured using selective laser melting. One way would be to use a support structure, but their removal is time- and cost-consuming. An alternative way is to use lattice structures to provide permanent support to all overhang areas of the body. Moreover, this lattice structure increased the stiffness of the component.

Finally, an analyze was run to find optimal parameters and to check the stiffness of the lightweight structure. Thanks to the capability of nTopology a change can be made at any point of the workflow and the entire workflow is updated automatically.

The last step of the design process was preparation for manufacturing. Additional material is required for post-processing to achieve high accuracy in some areas. For example, the holes were closed and to contact surfaces an offset of 0.5 mm was added.

3.2 Additive Manufacturing

To produce the Robo2Go head, the DMG MORI LASERTEC 30 DUAL SLM is used, which functions according to the operating principle of the SLM process. The maximum build volume of the system is 300×300 x 300 mm. Two 600 W lasers, each with scan fields covering the entire build volume, enable build volumes of up to 90cm3/h. In addition, the system is equipped with a permanent filter system that works independently of the material and, thanks to the automatic passivation of metal dust, offers a new level of durability and work safety. Active cooling of the installation space allows users to remove finished components earlier and thus increase productivity. All these features make the LASERTEC 30 DUAL SLM ideal to produce additive series components, like the robot head.

For manufacturing, the CAD model was exported from nTopology and imported into the preparation software. This is pardon of additive manufacturing to CAM programs in conventional manufacturing. The job preparation includes the position and orientation of the parts in the build space are defined, necessary overhangs of the components are made producible by support structures, and all machine parameters are defined. After the complete preparation, a machine file is generated. Based on this file, the additive manufacturing of the robot head took place.

The build volume of the DMG MORI LASERTEC 30 DUAL SLM allows the production of up to 4 robot heads made of AlSi10Mg simultaneously (Fig. 3). By using dual-laser configuration, the production time amounts to 39 h, instead of 73 h when using only single-laser configuration. So, the dual-laser configuration enables an 87% increase in productivity compared to a single-laser process.

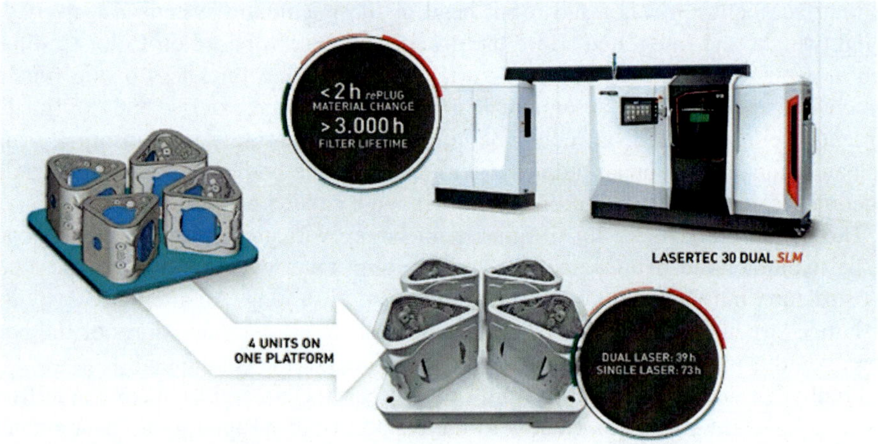

Fig. 3 Robo2Go—Additive Manufacturing Set-up

3.3 Post-Machining

Based on the process character of the additive production the post-machining itself is describable as subordinated process. The Robo2Go head passes through a total of six further steps before its finish (Fig. 4). At first, the lattice structures and the internal channels are thoroughly de-powdered using a wet separator (1). In connection the surface of the part is smoothened by particle blasting with glass beads (2). The two staged CNC milling process (3 and 5) are divided by the separation of the part and building plate using a conventional band saw (4). Last processing step is the addition of steel thread inserts to guarantee a rigid connection for the grippers while reduce the risk of damaging aluminum threads (6).

The CNC machining is the most extensive part, support material is removed, fittings are milled, threads are cut, and flat surfaces created. After complete deburring on the machine the finished part is taken out. For machining a custom fixture is responsible to hold the workpiece in position and allow 5 axis processing of all sides except the bottom. Required allowances for finishing are considered in the design phase as well as printed chamfers to decrease the machining effort. Due to small variations out of the building process the reference point for the final part was placed at the fixture to guarantee similar measurement and geometrical outcome for each component.

In the context of this paper, the post-processes enabled the usage of lightweight design and the optimization of the robot head. It is integral part of the process chain and while respecting its restrictions allows the usage of lattice structures and additive manufacturing in general. It underlines that AM is not replacing conventional production methods but used correctly empowers future manufacturing possibilities.

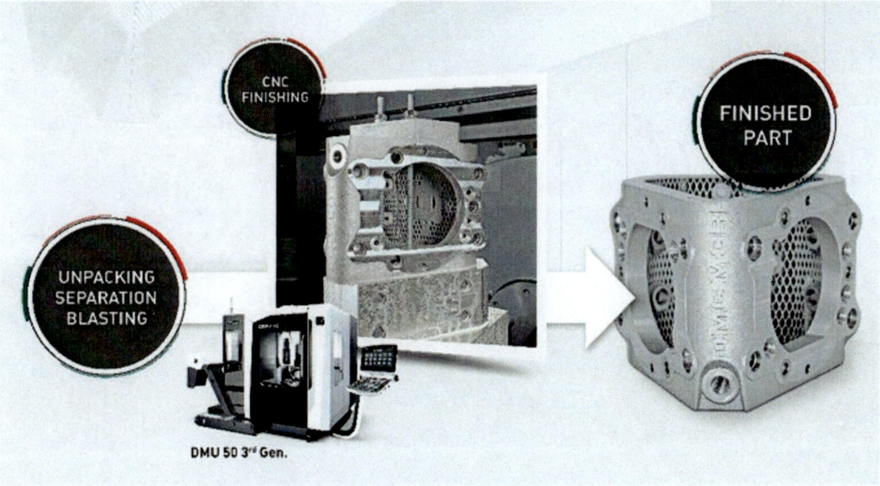

Fig. 4 Robo2Go—post-machining

4 Summary and Conclusion

The redesign of the Robo2Go head was the starting point in this project. In close collaboration with the customer, the requirements for the Robo2Go head and the available design space were defined. Afterwards a topology optimization was performed. The result was an organic design that resists the loads with minimal use of materials. In the last step, the customer's requirement of a closed housing was taken into account. The shell design has variable wall thicknesses that are based on the loads. The lattice structure on the inside allows a further reduction in weight with increased stiffness at the same time.

The advantages of additive manufacturing compared to conventional design can be seen clearly in the final design of the component. Besides the integration of the adapter plates, it was possible to integrate the required channels for the pneumatics without the otherwise error-prone sealing points. The new Robo2Go head is now fully compatible with all gripper types without any further adjustments. The key figures (Fig. 5) of the additively manufactured Robo2Go head impressively underline the potential of this technology. Thanks to the entire optimization from redesign to the process chain, positioning accuracy increased by the factor of 16.

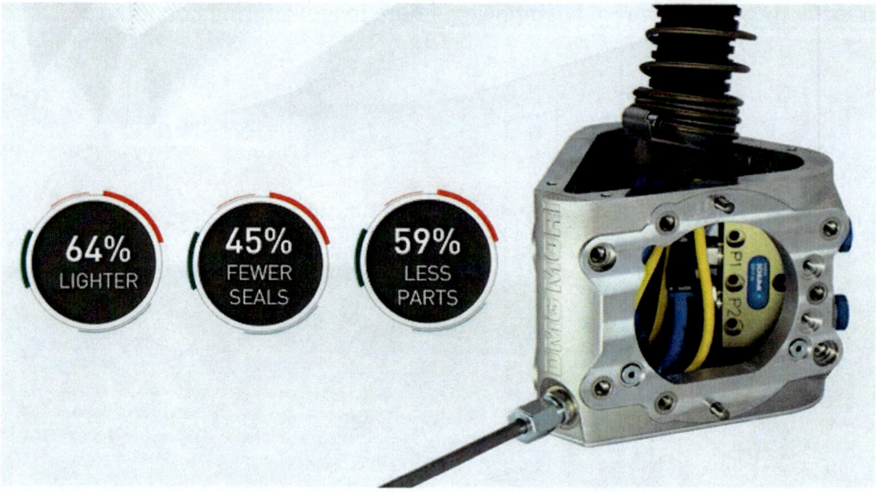

Fig. 5 Robo2Go—Summary of improvements

The new component has become 64% lighter due to the integrated lattice structures. At the same time, it was possible to achieve a higher stiffness of the structure due to the design. As a result of the integration of the pneumatic pipes into the base body, up to 45% of the previously needed sealing points are avoided. Thanks to the integral design, the number of individual parts is reduced by almost 60%. This consequently leads to a significant reduction in the effort in commissioning, assembly, and adjustment of the robot head. Furthermore, the center of gravity of the robot head was moved. It is now closer to the robot hand and results in an improvement of the system dynamics.

References

1. Gebhardt, A.: Additive Fertigungsverfahren. Additive Manufacturing und 3D-Drucken für Prototyping—Tooling -Produktion. Hanser, Munich (2016).
2. Klahn, C.; Meboldt, M..: Entwicklung und Konstruktion für die Additive Fertigung. Grundlagen und Methoden für den Einsatz in industriellen Endkundenprodukten. Vogel Business Media, Würzburg (2018).
3. Kumke, M.: Methodisches Konstruieren von additiv gefertigten Bauteilen. Dissertation (2018)

4. Zeyn, H.: Industrie 4.0 Industrialisierung der Additiven Fertigung“. Beuth Verlag, Berlin (2017).
5. Henning, F., Moeller, E.: Handbuch Leichtbau. Methoden, Werkstoffe, Fertigung. Hanser, Munich (2011).

A Unit Cell with Tailorable Negative Thermal Expansion Based On a Bolted Additively Manufactured Auxetic Mechanical Metamaterial Structure: Development and Investigation

Erhard Buchmann[1(✉)], Frank Hadwiger[2,3], Christoph Petroll[4], Christoph Zauner[3], Alexander Horoschenkoff[2], and Philipp Höfer[1]

[1] Institute of Lightweight Engineering, Universität der Bundeswehr München, Neubiberg, Germany
{erhard.buchmann,philipp.hoefer}@unibw.de
[2] Munich University of Applied Sciences, Munich, Germany
alexander.horoschenkoff@hm.edu
[3] KRP Mechatec GmbH, Garching, Germany
hadwiger@krp-m.de, christoph.zauner@largespace.de
[4] Wehrwissenschaftliches Institut für Werk- und Betriebsstoffe, Erding, Germany
christophpetroll@bundeswehr.org

Abstract. There is a growing interest in the application of mechanical metamaterials due to the recent advances in additive manufacturing technology. In particular, materials with adjustable thermal expansion have many fields of utilization in engineering. Materials with negative thermal expansion (NTE) can be used in combination with materials with positive thermal expansion for creating thermoelastically stable structures with an ultra low coefficient of thermal expansion (CTE). NTE-lattice structures generally require multi-material combinations to achieve the desired CTE. However, multi-material 3D printing is currently in development and not available for industrial-scale applications to date. In this paper, we present a unit cell based on an auxetic mechanical metamaterial structure that can be manufactured using single material additive manufacturing. For investigation, unit cells and a unit cell tessellation with certain CTEs were designed and manufactured using metallic materials. The mechanical and thermoelastic functionality of the designed unit cell could be demonstrated experimentally regarding the CTE and the stiffness. The presented approach for including cells with tuneable NTE and stiffness in additively manufactured structures has a high potential for realization in practice.

J. Rieser et al. (Eds.): *Proceedings of the Munich Symposium on Lightweight Design 2021*, Proceedings, pp. 198–211, 2023. https://doi.org/10.1007/978-3-662-65216-9_18

Keywords: Metamaterial · Smart materials · Negative thermal expansion · Additive manufacturing · Auxetic · Unit cell · Thermoelasticity

1 Introduction

Mechanical metamaterials have a rationally designed artificial micro structure that enables certain desired properties on macroscale [1]. Usually, the micro structure shows a unit cell that is composed periodically in order to create a macro scale structure. Due to the recent advances in additive manufacturing (AM) technologies, it is more and more possible to manufacture arbitrary complex geometries on several length scales for many materials which further advance this field of research [2,3]. The properties of mechanical metamaterials encompass, but are not limited to, ultralight and ultrastiff structures [4], structures with a negative Poisson's ratio [5], and structures with a designed deformation pattern that facilitates building mechanisms [6]. This enables a wide range of application in engineering and science [3]. One class of metamaterials are thermoelastic metamaterials that have a designed coefficient of thermal expansion. On the one hand, they can serve as actuator [7,8], on the other hand they can be used to design structures with a certain thermoelastic behavior [9,10]. In particular, metamaterials with negative thermal expansion (NTE) can be used in combination with materials with positive thermal expansion for creating thermoelastically stable structures with ultra low coefficient of thermal expansion (CTE).

For the creation of NTE-metamaterials at least two components with different positive CTEs are required [11]. In order to achieve the desired overall thermal expansion, those constituents are arranged in special patterns, of which many have been proposed [12]. Many of these concepts have been tested using polymeric materials. But, for many applications, as an example for space applications, metallic materials are more suitable due to the higher temperature application range, better outgassing properties, and the superior stiffness.

Regarding the application of additive manufacturing, a major challenge is the joining of the two constituents. Multi-material additive manufacturing is currently in development and not available for industrial-scale applications to date. Interestingly, Ding et al. [13] fabricated a metallic metamaterial with NTE using a robotized laser powder-feed metal AM-system. As the authors were focusing on the manufacturing process, the functionality of the structure, especially the stiffness and the CTE were not examined. Parsons [14] investigated a NTE metamaterial structure of aluminum and titanium. He applied ultrasonic additive manufacturing to build a block structure with titanium and aluminum bands. This AM method has many limitation regarding the geometric complexity which was not further studied. The metamaterial structure was then cut out of this block. However, it is not addressed how the structure could be manufactured directly without an extra cutting process. Moreover, Parsons outlines different methods for manually joining (pins, bolts, interference fits, adhesives, soldering, brazing or welding) applied in previous research work. Steeves et al. [15] investigated conventionally manufactured samples joined with interference fits, pins and bolts. These approaches of previous research work generally require the rather complex conventional manufacturing of two parts of different materials that are then joined. However, most of the studies to date focus on the investigation of underlying mechanisms of thermoelastic metamaterials. The design freedom when using additive manufacturing is generally not adressed. In particular, only limited studies show how to integrate thermoelastic metamaterials into components in practice. Moreover, the possibilities of AM for reducing post-processing as cutting and joining effort are not considered.

In this work, we present a unit cell with tailorable NTE based on an auxetic mechanical metamaterial structure that can be manufactured using single material AM and assembled straightforwardly using a customary screw. For investigation, unit cells and an unit cell tessellation with certain CTEs were designed and manufactured using aluminum. The functionality is then examined with tests regarding the stiffness and the CTE.

2 Design of the Unit Cells

2.1 Thermal Expansion Mechanism of the Unit Cell

The concept proposed herein is based on the auxetic planar cellular honeycomb structure, that was studied in [16–18] (see Fig. 1a).

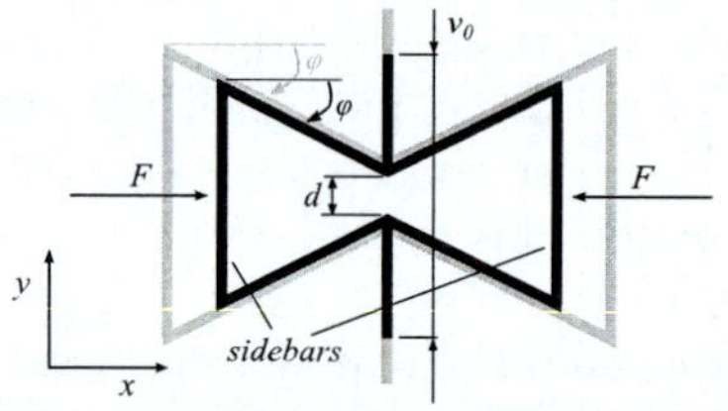

(a) Auxetic honeycomb structure. Initial state (black) and heated state (grey).

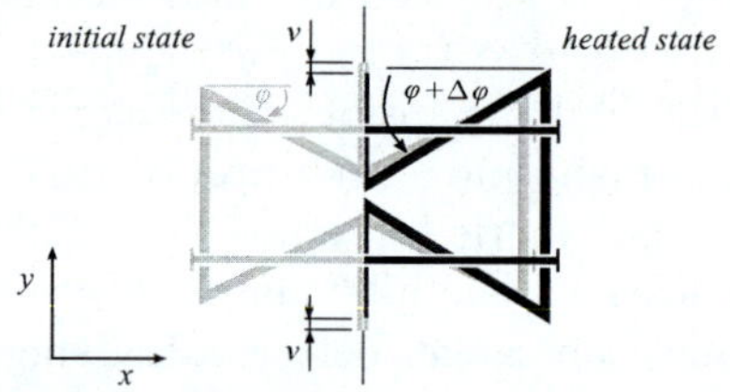

(b) Bolted auxetic honeycomb structure. Inital state (grey) and heated state (black, half model).

Fig. 1. Auxetic honeycomb structure without (a) and with (b) bolt showing the thermoelastic kinematics caused by thermal expansion

The re-entrant mechanism ensures the reduction of width v_0 when stressed by the force F perpendicularly by a small increase of the angle φ. A uniform heating of the cell results in scaling of the structure without changing the angle φ. In this work, the thermal expansion of the unit cell in x-direction is hindered by using a bolt with a smaller CTE than the auxetic structure (see Fig. 1b). Due to the re-entrant mechanism, this results in a reduction of v_0 by $2v$ in the heated state.

So, a thermal expansion smaller than the CTE of the material of the auxetic structure y-direction, including NTEs can be realized. Simultaneously, the unit cell expands through the sidebars in y-direction. Consequently, a small gap d makes the thermal expansion mechanism efficient when aiming for NTEs. When subjected to temperatures under the installation temperature, the bolt must be preloaded to ensure the function of the re-entrant mechanism. This unit cell can be integrated into an arbitrary complex AM geometry for a local thermoelastic functionalization. By using a customary screw with nut as bolt, the structure can be assembled straightforwardly after 3D printing without extra manufacturing effort. The two main properties regarding the application, the stiffness and the CTE, substantially depend on the angle φ. The CTE was evaluated analytically using a quarter model which is outlined in the following.

2.2 Calculation of Thermoelastic Kinematics Caused by Thermal Expansion

For the calculation of the kinematics of thermal expansion an idealized truss model of the unit cell is considered. The truss consists of beams that are connected with ideal joints, whereby the joint stiffness is modeled with torsion springs (see Fig. 2a). In this course, the shear deformation and the bending deformation of the beams is neglected due to the low stiffness of the torsion springs in the experimental structures (see Sect. 2.3). The positions of the joints in the model refers to the experimental structures (see Fig. 4a). In Table 1 the main model parameters are listed.

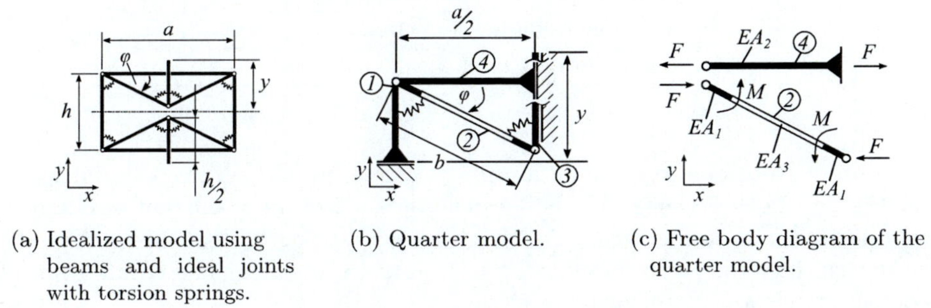

(a) Idealized model using beams and ideal joints with torsion springs.

(b) Quarter model.

(c) Free body diagram of the quarter model.

Fig. 2. Models used for the analytical calculation of the thermoelastic kinematics caused by thermal expansion. Beam 2 has two axial stiffnesses, EA_1 (black sectors) and EA_3 (white sector). Beam 4 has the axial stiffness EA_2

Due to the symmetry, the study of a quarter model is adequate (see Fig. 2b). For the beams two coefficients of thermal expansion α are regarded, α_1 (beams $1, 2, 3$) and α_2 (beam 4), with $\alpha_1 > \alpha_2$. For the calculation all beams are subjected to the same temperature load ΔT and are subjected to the reaction forces in the joints. Consequently, the total expansion of a beam of the length l is the sum of two parts. On the one hand the thermal expansion Δl_{th}, on the other hand the mechanical expansion through the reaction forces Δl_m. The thermal expansion Δl_{th} is considered as $\Delta l_{th} = l \cdot \alpha \cdot \Delta T$. For the mechanical expansion Δl_m through the force F the linear formula $\Delta l_m = l \cdot \frac{F}{EA}$ using the axial stiffness EA is applied. Two axial stiffnesses for beam 2 are considered (see Fig. 2). EA_1 is the axial stiffness of the hinges whereas EA_3 is the axial stiffness of the solid part of beam 2. The axial stiffness of the solid part is far higher than the axial stiffness of the hinges at the experimental structures in this work (see Sect. 2.3). Therefore, the solid part is seen as rigid here.

Table 1. Main model parameters for the calculation of the thermoelastic kinematics caused by thermal expansion

Symbol	Description
Δl_{th}	Thermal expansion
Δl_m	Mechanical expansion
α_1	Coefficient of thermal expansion of beams $1, 2, 3$
α_2	Coefficient of thermal expansion of beam 4
EA_1	Axial stiffness of the hinges adjacent to beam 2
EA_2	Axial stiffness of beam 4
EA_3	Axial stiffness of beam 2
C	Joint stiffness
a	Unit cell dimension in x-direction
b	Length of re-entrant mechanism leg
h	Unit cell dimension in y-direction
φ	Angle of re-entrant mechanism

In order to calculate the reaction forces a free body diagram is investigated (see Fig. 2c). In this course, the change of the angel φ of the re-entrant mechanism when heated is described with $\Delta\varphi$. Each torsion spring with stiffness C at the joints then introduce the moment $M = C \cdot \Delta\varphi$ on beam 2. The equilibrium of moments for beam 2 of length b gives

$$F = \frac{2 \cdot C \cdot \Delta\varphi}{b \cdot \sin(\varphi + \Delta\varphi)}. \tag{1}$$

For the total expansion in y-direction, the geometric compatibility of beam 2 and 4 must be met:

$$\frac{a}{2} + \Delta l_{th_4} + \Delta l_{m_4} = (b + \Delta l_{th_2} + \Delta l_{m_2}) \cdot \cos(\varphi + \Delta\varphi) \tag{2}$$

When using the formulas for Δl_{th}, Δl_m together with Eqs. (1) and (2), the angular change $\Delta\varphi$ and the reaction force F can be calculated.

In order to determine the CTE of the unit cell the section y in undeformed state (y_0) and in heated state (y_{th}) is calculated. The sections y_0 and y_{th} can be described as follows.

$$y_0 = h - \frac{a}{2} \cdot \tan\left(\arccos\left(\frac{a}{2 \cdot b}\right)\right) \tag{3}$$

$$y_{th} = h + \alpha_1 \cdot h \cdot \Delta T - \tan(\varphi + \Delta\varphi)\left[\left(\frac{a}{2} + \alpha_2 \cdot \frac{a}{2} \cdot \Delta T + \frac{F}{EA_1} \cdot \frac{a}{2}\right)\right] \tag{4}$$

This gives for the CTE in y-direction of the whole unit cell

$$CTE_y(a, b, h, \alpha_1, \alpha_2, \varphi, \Delta T, EA_1, EA_2, C) = \frac{y_{th} - y_0}{y_0 \cdot \Delta T} \tag{5}$$

The CTE in x-direction is the same as for beam 4.

$$CTE_x(a, b, h, \alpha_1, \alpha_2, \varphi, \Delta T, EA_1, EA_2, C) = \frac{\Delta l_{m_4} + \Delta l_{th_4}}{\frac{a}{2} \cdot \Delta T} = \frac{F}{EA_2 \cdot \Delta T} + \alpha_2 \tag{6}$$

Notably, beams 2 and 4 are stretched or compressed respectively during heating depending on the ratio of the axial stiffnesses and the joint stiffness. In the following, the governing equations shall be used to design the experimental unit cells and the unit cell tessellation.

2.3 Design of Experimental Unit Cells

Equations (5) and (6) show that the CTEs can be influenced by many material and geometric parameters. For the selection of materials see Sect. 3.1. In order to provide adequate reference surfaces for the capacitive displacement sensors, a depth of 15 mm was chosen (see Table 2 and Fig. 4b).

To guarantee the system kinematics, hinges were used as joints. Hinges can be manufactured directly and need no further mounting procedure. As thickness of the hinges 0.5 mm was used. Preliminary experiments showed that these hinges can be manufactured robustly and provide enough flexibility. Hence, the beams were designed to 3 mm, to have a significant higher cross section of the beams compared to the hinge. The hinge stiffness C was determined experimentally (3000 Nmm rad^{-1}). The expansion mechanism requires a hole in beam 2 for the bolt (see Fig. 4a). For the calculation, the influence of the hole on the axial stiffness on beam 2 is neglected. Moreover, as to prevent the hole passing through the hinges, the angle must be reasonably large.

This work focuses on studying the influence of the angle φ. In order to make the unit cells comparable, a constant width a of 60 mm and a constant gap d of 1 mm was utilized. Consequently, b and h are depended parameters that can be calculated as follows.

$$b = \frac{a}{2} \cdot cos(\varphi) \tag{7}$$

$$h = a \cdot \sin(\varphi) + 1 \text{ mm} \tag{8}$$

Table 2. Main geometric dimensions for the three investigated unit cells (UC+3, UC-16, UC-71)

Parameter type	Value	Unit	UC+3	UC-16	UC-71
Design target	CTE_y	$10^{-6}\,K^{-1}$	2.7	−15.8	−70.6
Variable	φ	°	37.6	28.3	18.5
	a	mm	60		
	d	mm	1		
Constant	C	Nmm rad^{-1}	3000		
For all unit cells	Depth of models	mm	15		
	Thickness of beams	mm	3		
	Thickness of hinges	mm	0.5		
	b	mm	37.9	34.1	31.6
Dependent	h	mm	47.2	33.3	21.1
	CTE_x	$10^{-6}\,K^{-1}$	9.7	9.7	9.7

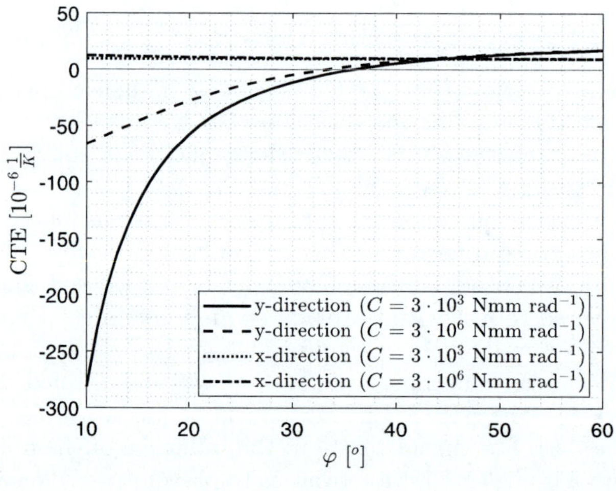

Fig. 3. Influence of the angle φ on the CTEs using the values of Tables 2 and 3. While CTE_x stays almost constant at approx. $9.7 \cdot 10^{-6}$ K^{-1} for all φ, the CTE_y asymptotically approximates the CTE of the auxetic base material with rising φ. For decreasing φ the CTE_y gets more sensitive to changes of φ. Negative CTEs in y-direction can be realized with angles smaller than approx. 36°. The influence of the torsion stiffness C is very small here. Only when using a value far higher ($C = 3 \cdot 10^6$ Nmm rad^{-1}) instead of $C = 3 \cdot 10^3$ Nmm rad^{-1} a notable deviation can be seen

The target coefficient of thermal expansion CTE_y can now be designed using the angle φ (see Fig. 3).

CTE_x stays almost constant at approximately $9.7 \cdot 10^{-6}$ K^{-1} for all φ. The CTE_y asymptotically approximates the CTE of the auxetic base material with rising φ. Notably, all the designed CTE_y are smaller than those of the auxetic base material. For decreasing φ, the CTE_y gets more sensitive to changes of φ. Negative $CTEs$ in y-direction can be realized with angles smaller than approximately 36°.

The influence of the torsion stiffness C is very small for the structure investigated here. Only when using a value far higher a notable deviation can be seen (see Fig. 3). In this case the thermal expansions of the beams Δl_{th} stay the same while the mechanical expansions get higher due to the higher reaction forces (see Eq. 1). This results in a more positive CTE in x and y-direction. Notably, the mechanism is stressfree for any thermal state if $C = 0$ is applied (see Eq. 1).

For the experimental test, unit cells (UC) with CTE_y of $+2.7 \cdot 10^{-6}$ K^{-1} (UC+3), $-15.8 \cdot 10^{-6}$ K^{-1} (UC-16, see Fig. 4a) and $-70.6 \cdot 10^{-6}$ K^{-1} (UC-71) were designed (see Table 2). As the mechanical expansion is very small, the CTE_x of $9.7 \cdot 10^{-6}$ K^{-1} differs insignificantly from the one of titanium. Moreover, a 3x3 tessellation structure of the UC-16 unit cells (UCM-16) was investigated (see Fig. 4c).

(a) Unit cell UC+3 with grinded surfaces for the measurement rig with displacement sensors (1) and the titanium screws (2). The calculation model (3) refers to the center lines.

(b) Unit cell UC+3 with measurement rig, displacement and temperature sensors.

(c) Matrix unit cell structre (UCM-16).

Fig. 4. Investigated unit-cell UC+3 without (a) and with measurement equipment (b). Unit cell tessellation structure (UCM-16) (c). In (a) the deviation of the position of the joints in the calculation model and the position of the hinges of the printed structure can be seen

3 Materials, Manufacturing and Experimental Procedure

3.1 Materials

An aluminum alloy was used for the AM structure as it is a common material for structural applications. Moreover, additive manufacturing facilities for aluminum alloys are wide spread in industry. For the bolt a M3 titanium threaded rod with nuts and washers was used. The combination of titanium and aluminium forms a notable CTE gap what allows the effective creation of the thermal expansion properties of the metamaterial. For the determination of the coefficients of thermal expansion of both metals a thermomechanical analysis (TMA) was applied.

Table 3. Material data for analysis. Mean CTEs for temperatures between 20 °C and 135 °C were measured using TMA

Material	CTE (TMA)	Young's modulus [19]
Aluminum (AlSi10Mg)	$21.1 \cdot 10^{-6}\mathrm{K}^{-1}$	69 GPa
Titanium (Grade 2)	$9.7 \cdot 10^{-6}\mathrm{K}^{-1}$	110 GPa

3.2 Manufacturing and Experimental Procedure

For the unit cells and the unit cell tessellation selective laser sintering (SLS) parts were manufactured. In order to avoid support structure in the AM process, the holes for the titanium bolt in the auxetic base structure were designed elliptic. The M3 titanium threaded rods were mounted to the manufactured structure with a torque of <1 Nm using a high-precision torque wrench. This preload was enough to avoid a settling effect during heat up and to limit the elastic deformation through the preload to a minimum. The auxetic part has functional surfaces for placing the measurement rig with displacement sensors and the titanium threaded rod with washer which is depicted in Fig. 4a. All functional surfaces are protruded and could be grinded after the additive manufacturing.

For the examination of the mechanical and thermoelastic functionality two types of experiments were conducted. (i) The stiffness in y-direction was evaluated using a common tensile testing machine. With the actual design, the unit cell can only be loaded by traction. (ii) The thermoelastic characterization test was performed in a heating chamber with circulating air. The forced convection ensures a nearly homogeneous heat up of all areas of the specimen. Two cycles from room temperature to 135 °C were performed. The temperature was measured redundant at the surface using two thermocouples. The deformations of the unit cell in x-direction and y-direction were measured contact-free with capacitive displacement sensors. The displacement sensors have a measurement accuracy of ±0.03%. Measurement rigs were manufactured during the same printing job than the samples. The rig is mounted on one side to the mounting reference surface of the specimen and on the other side the displacement sensor is applied. The temperature dependent CTE of the rig is well known by a TMA measurement and can be compensated afterwards. Hence, the expansion of the specimens can be measured suitable. The instantaneous CTE was evaluated for each measurement according to ASTM-E228-17 [20].

4 Experimental Results and Discussion

4.1 Test of the Biaxial Thermal Expansion Behavior

The measured mean CTEs in x and y-direction were compared to the calculated CTEs which is depicted in Fig. 5. Generally, there is a good concordance of the calculated values of CTE_y to the measured ones regarding the unit cells. Consequently, the concept is appropriate for designing unit cells with positive CTEs smaller than the auxetic base material to unit cells with a high NTE. The concurrence for the unit cells UC+3 and UC-16 with a deviation of 0.6 µm and 1.7 µm respectively is better compared to the UC-71 with a deviation of 9.2 µm. Hence, there is a higher deviation between the measured CTE_y compared to the calculated values at lower values of the CTE. In this work, lower values of the CTE are reached with lower angles φ. This is in accordance with the angular sensitivity of the unit cells. The unit cell is more sensitive for changes in φ for lower values of φ, which can be seen in Fig. 3. Consequently, the discrepancies in the angle φ between the calculation model and the experimental structure result in higher deviations of the measured CTEs compared to the calculated ones at lower angels φ. Regarding the inclination of CTE_y with respect to φ, the change in sensitivity is nonlinear and much higher at angles lower than approximately 20° (see Fig. 3). Therefore, the deviations of CTE_y are comparable at UC+3 and UC-16, and much higher at UC-71.

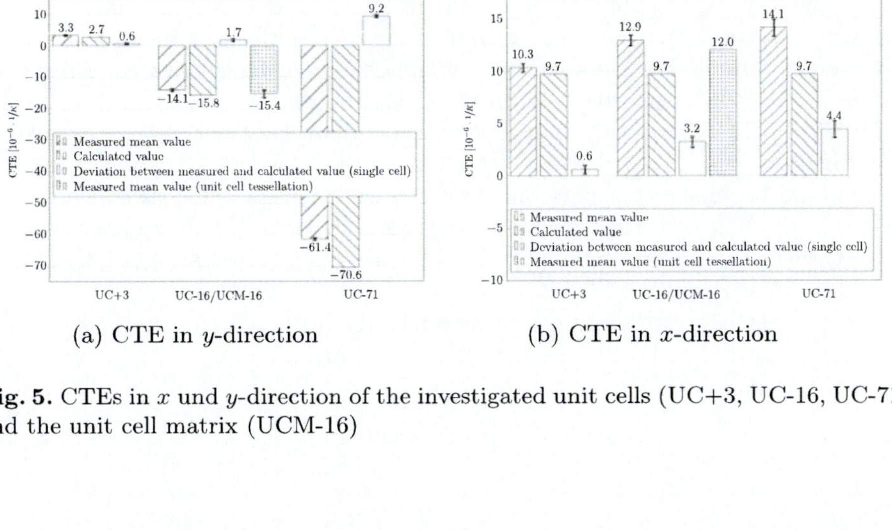

(a) CTE in y-direction (b) CTE in x-direction

Fig. 5. CTEs in x und y-direction of the investigated unit cells (UC+3, UC-16, UC-71) and the unit cell matrix (UCM-16)

The deviation of CTE_x rises with lower angles or lower CTEs respectively, too. While the deviation of CTE_x at UC-71 is 4.4 μm, it is lower for UC-16 (3.2 μm) and for UC+3 (0.6 μm). Regarding the unit cell tessellation, the deviation of CTE_y is higher compared to the unit cell, while the deviation of CTE_x is lower.

It is assumed that there are two main reasons for the deviations of the measured values. First, there are inaccuracies related to the joining of the titanium threaded rods. The holes in the auxetic structure for the bolts provide a very loose fit to the titanium threaded rods. Therefore, the two rods could be assembled non-parallely. Secondly, there are geometric deviations of the calculation model compared to the printed structure (see Fig. 4a). The titanium screw is larger than in the calculation model caused by the thickness and the protrusion on the aluminum beams. Therefore, the expansion of the unit cell in x-direction is higher, which is in accordance with the measured values of CTE_x (see Fig. 5b). Moreover, this lowers the re-entrant effect of the mechanism. This could contribute to the lower measured NTE values of CTE_y of UC-71 and UC-16. Also, this matches with the lower deviation of CTE_x and CTE_y of the unit cell tessellation. There, the thickness of the aluminum truss and the protrusion contribute only two times for three unit cells. Therefore, the overlength of the titanium threaded rod with respect to the calculation model is shorter per unit cell compared to the single cell model.

Another point is the deviation of the joint position between the calculation model and the printed structure. At the printed structure, the length h is shorter which lowers the CTE in y-direction. This is in accordance with the lower measured CTE compared to the calculated one of UC+3. Moreover, the truss b is shorter, which results in a shorter lever arm of the re-entrant mechanism and therefore a reduction of the NTE-effect in y-direction. This is in coincidence with UC-16 and UC-71.

During the measurement a strong dependence of the CTEs of the temperature was observed, which is depicted exemplary for UC-16 in Fig. 6a. Therefore, a temperature-dependent calculation of the CTEs of the unit cell UC-16 was applied using the temperature-dependent CTEs of aluminum and titanium measured in the TMA. Both measured CTEs of the unit cell increase with rising temperature rather linearly which is in good accordance with the calculated CTEs. In concordance with the mean CTEs the deviation of CTE_x is bigger than CTE_y. In future, the temperature-dependent behavior should be further investigated.

It could be shown that with this unit cell wide ranges of CTEs and NTE can be realized. Hence, the unit cell can serve as a thermal actuator or as designed compensation actuator for a structures with a longer characteristic length than the unit cell with positive CTE.

In the experiments the unit cells were only investigated over the installation temperature, as the mechanism does not work at lower temperatures with the small preload. This can be overcome by assembling the mechanism under the operating temperatures, by using higher preloads or by locking the axial movement of the titanium rods in both directions.

As can be seen in Eq. (5) there are many possibilities to increase the CTE efficiency of the unit cell. Regarding Eq. (4), one possibility for improving the NTE-effect would be to lower α_2. On the one hand, this could be done using a carbon fiber as rod instead of the titanium threaded rod. On the other hand, a rod made of shape memory alloy could be utilized to further increase the NTE-effect. As the reaction forces are rather low, rods with far lower axial stiffness, e.g. wires, are suitable as well, when only stiffness for tensile loads on the unit cell is required. This would further broaden the possibility for integrating this unit cell into components. After the manufacturing of a component with several unit cells in arbitrary positions the wire could be run in afterwards through all cells serving as rod.

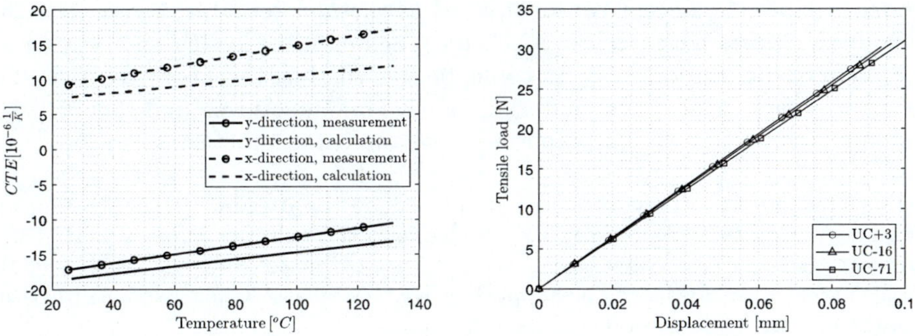

(a) (b)

Fig. 6. Temperature-dependent CTEs of the UC-16 unit cell (a) and stiffness in *y*-direction of UC+3, UC-16, UC-71 (b)

4.2 Test of Stiffness in the Direction of Thermal Expansion

Besides the kinematics of thermal expansion, the mechanical properties have to be considered when applying the unit cells into components. In general, the CTEs and the stiffness of the unit cell depend on each other (see Sect. 2.2). Notably, cells with lower NTE have smaller angles φ. This results in higher reaction forces under thermal loads, which can be deduced from Eq. (2). But, because of the low torsion stiffness, the reaction forces are very small (see Eq. (1)). Consequently, the mechanical elongation of the trusses under thermal loads can be neglected here.

Regarding the mechanical properties, the stiffness of the unit cells in the direction of thermal expansion was tested experimentally. The tensile test of the three unit cells showed a similar stiffness value and a similar linear behavior of the tensile load with respect to the deformation (see Fig. 6b). In this course, the stiffness slightly decreases with higher angels. This behavior can be explained with an increasing shear deformation of the hinges at lower angles.

5 Conclusion

In this paper we designed and investigated a unit cell with tailorable CTE based on a bolted additively manufactured auxetic mechanical metamaterial structure. First, the thermoelastic behaviour was studied analytically to predict the thermal expansion for the designed unit cells. Then the thermoelastic functionality was assessed experimentally for three different unit cells and an unit tessellation. Additionally, the stiffnesses of the respective unit cells were evaluated by a mechanical loading test. A good concordance between the calculated and measured CTEs was found in a wide range of CTEs, including a low thermal expansion ($3.3 \cdot 10^{-6}$ K^{-1}) and a highly negative CTE of $-61.4 \cdot 10^{-6}$ K^{-1}. Hence, the presented unit cell concept can be applied as thermal actuator or compensation unit for structures with positive CTE for reaching ultra low thermal expansion of a component. Importantly, it could be shown that the presented manufacturing approach using single material additive manufacturing and then fitting a customary screw has a high potential to be applied in practice.

Acknowledgments. This work was supported by the dtec.bw—Digitalization and Technology Research Center of the Bundeswehr through the Project SeRANIS—Seamless Radio Access Networks in the Internet of Space under Grant 150009910. We also greatly acknowledge the support of KRP Mechatec GmbH and Wehrwissenschaftliches Institut für Werk- und Betriebsstoffe.

References

1. A.A. Zadpoor, Materials Horizons **3**(5), 371 (2016)
2. M. Askari, D.A. Hutchins, P.J. Thomas, L. Astolfi, R.L. Watson, M. Abdi, M. Ricci, S. Laureti, L. Nie, S. Freear, et al., Additive Manufacturing p. 101562 (2020)
3. J.U. Surjadi, L. Gao, H. Du, X. Li, X. Xiong, N.X. Fang, Y. Lu, Advanced Engineering Materials **21**(3), 1800864 (2019)
4. X. Zheng, H. Lee, T.H. Weisgraber, M. Shusteff, J. DeOtte, E.B. Duoss, J.D. Kuntz, M.M. Biener, Q. Ge, J.A. Jackson, et al., Science **344**(6190), 1373 (2014)
5. A. Joseph, V. Mahesh, D. Harursampath, Advances in Manufacturing **9**(3), 342 (2021)
6. A. Ion, J. Frohnhofen, L. Wall, R. Kovacs, M. Alistar, J. Lindsay, P. Lopes, H.T. Chen, P. Baudisch, in *Proceedings of the 29th annual symposium on user interface software and technology* (2016), pp. 529–539
7. O. Sigmund, S. Torquato, Applied Physics Letters **69**(21), 3203 (1996)
8. M.M. Toropova, C.A. Steeves, in *Smart Materials, Adaptive Structures and Intelligent Systems*, vol. 57298 (American Society of Mechanical Engineers, 2015), vol. 57298, p. V001T01A002
9. N. Yamamoto, E. Gdoutos, R. Toda, V. White, H. Manohara, C. Daraio, Advanced Materials **26**(19), 3076 (2014)
10. M.M. Toropova, C.A. Steeves, Acta Astronautica **113**, 132 (2015)

11. A.L. Kalamkarov, A.G. Kolpakov, *Analysis, design and optimization of composite structures*, vol. 1 (Wiley New York, 1997)
12. H. Xu, A. Farag, D. Pasini, Journal of the Mechanics and Physics of Solids **117**, 54 (2018)
13. Y. Ding, M. Akbari, X.L. Gao, L. Ai, R. Kovacevic, in *Manufacturing techniques for materials: engineering and engineered* (CRC Press, 2018), pp. 51–65
14. E.M. Parsons, Composite Structures **223**, 110656 (2019)
15. C.A. Steeves, S.L.d.S. e Lucato, M. He, E. Antinucci, J.W. Hutchinson, A.G. Evans, Journal of the Mechanics and Physics of Solids **55**(9), 1803 (2007)
16. L.J. Gibson, M.F. Ashby, G. Schajer, C. Robertson, Proceedings of the Royal Society of London. A. Mathematical and Physical Sciences **382**(1782), 25 (1982)
17. R. Almgren, J. Elast **15**, 427 (1985)
18. T.L. Warren, Journal of applied physics **67**(12), 7591 (1990)
19. D.A. de Rooij. Spacematdb - space material database. https://www.spacematdb.com/spacemat/
20. ASTM International, Standard test method for linear thermal expansion of solid materials with a push-rod dilatometer. Standard ASTM E228-17, West Conshohocken, PA (2017)

Author Index

J. Rieser et al. (Eds.): *Proceedings of the Munich Symposium on Lightweight Design 2021*, Proceedings, pp. 213–214, 2023. https://doi.org/10.1007/978-3-662-65216-9